MW00898209

MATH REFRESHER FOR ADULTS

The Complete Guide to Boost your Math Skills Quickly with a Step-by-Step Framework and Hands-On Exercises (Bonus Workbook Included)

Alex Carter & BrightMind Press

© Copyright 2024 - All rights reserved.

The content contained within this book may not be reproduced, duplicated or transmitted without direct written permission from the author or the publisher.

Under no circumstances will any blame or legal responsibility be held against the publisher, or author, for any damages, reparation, or monetary loss due to the information contained within this book, either directly or indirectly.

Legal Notice:

This book is copyright protected. It is only for personal use. You cannot amend, distribute, sell, use, quote or paraphrase any part, or the content within this book, without the consent of the author or publisher.

Disclaimer Notice:

Please note the information contained within this document is for educational and entertainment purposes only. All effort has been executed to present accurate, up to date, reliable, complete information. No warranties of any kind are declared or implied. Readers acknowledge that the author is not engaged in the rendering of legal, financial, medical or professional advice. The content within this book has been derived from various sources. Please consult a licensed professional before attempting any techniques outlined in this book.

By reading this document, the reader agrees that under no circumstances is the author responsible for any losses, direct or indirect, that are incurred as a result of the use of the information contained within this document, including, but not limited to, errors, omissions, or inaccuracies.

Table of Contents

Introduction vii

CHAPTER 1: Numbers and Operations 1

Natural, Whole, Integer, Rational, and Irrational Numbers 1

Order of Operations (PEMDAS) 4

CHAPTER 2: Fractions and Decimals 8

Converting Between Fractions and Decimals 8

Arithmetic Operations With Fractions 10

CHAPTER 3: Percentages and Ratios 16

Percentage Increase and Decrease Calculations 16

Ratio Problem-Solving Techniques 22

CHAPTER 4: Factors and Multiples 26

Prime Factorization Methods 26

Least Common Multiple Calculations 29

CHAPTER 5: Basic Algebraic Concepts 33

Simplifying Algebraic Expressions 33

Recognizing and Combining Like Terms 35

CHAPTER 6: Working With Integers and Exponents 38

Integer Arithmetic Operations 38

Exponentiation Rules 43

CHAPTER 7: Understanding Proportions and Rates 47

Proportion Problem Strategies 47

Diagnostic Quizzes on Proportions and Rates 51

Reflection Questions on Pre-Algebra Essentials 51

CHAPTER 8: Geometry Fundamentals..**53**

Different Types of Angles 54

Measurement Techniques in Geometry 55

CHAPTER 9: Shapes and Their Properties.........................**60**

Perimeter and Area of 2D Shapes..61

Volume Calculations for 3D Shapes..64

Reflection Questions on Geometry Basics 67

CHAPTER 10: Working With Algebraic Expressions **69**

Factoring Algebraic Expressions 69

Simplifying Polynomials 73

CHAPTER 11: Solving Linear Equations and Inequalities **76**

Graphing Solutions on a Number Line 76

Real-World Applications of Equations 78

CHAPTER 12: Introduction to Graphing **81**

Plotting Linear Equations..82

Understanding and Calculating Slope 83

Reflection Questions on Intermediate Algebra 86

CHAPTER 13: Financial Math **88**

Simple and Compound Interest Calculations 88

Personal Budgeting Techniques 91

CHAPTER 14: Statistics and Probability...........................**95**

Calculating Mean, Median, and Mode 95

Basic Probability Concepts...98

Reflection Questions on Real-Life Math Applications 100

CHAPTER 15: Comprehensive Review **102**

Problem Sets Focused on Difficult Areas...................................... 102

Timed Drills for Speed and Accuracy 104

Suggested Tools 106

CHAPTER 16: Final Practice Test...**108**

Practice Test 109

Conclusion **113**

Solutions **116**

Numbers and Operations 116

Fractions and Decimals.. 116

Percentages and Ratios 117

Factors and Multiples 117

Reflection Questions on Arithmetic Foundations 118

Basic Algebraic Problems 119

Integers and Exponents.. 119

Proportions and Rates 120

Reflection Questions on Pre-Algebra Essentials 120

Geometry Fundamentals.. 121

Shapes and Their Properties... 121

Reflection Questions on Geometry Basics 121

Algebraic Expressions.. 122

Linear Equations and Inequalities.. 122

Introduction to Graphing 122

Reflection Questions on Intermediate Algebra.................................... 122

Financial Math.. 123

Statistics and Probabilities 123

Reflection on Real-Life Math Applications 124

Practice Test 124

References **131**

Introduction

Math is an integral part of our daily lives, yet many of us have struggled with it at some point or another. Say you're sitting at the kitchen table, your child's math homework spread out in front of you. You stare at the problem, trying to remember how to simplify fractions, but all you feel is a creeping sense of dread. Suddenly, memories of struggling through math classes in school flood back, and you wonder how you'll ever be able to help your child succeed. If this scenario sounds familiar, know that you're not alone.

Many adults face similar challenges when it comes to math. Whether it's assisting children with their homework, preparing for college entrance exams, or needing a refresher for career advancement, math tends to come with its own set of anxieties and hurdles. This book is here to change that. Designed as a user-friendly guide, it aims to help you rebuild your confidence in math by offering clear explanations, practical examples, and a structured approach tailored to meet the unique needs of adult learners.

One of the key purposes of this book is to strip away the fear and stress often associated with math. We understand that starting or returning to learning can be intimidating, especially if you feel out of practice. But don't worry—we've got your back. This book is crafted to make math approachable and manageable, breaking down complex concepts into easy-to-understand sections so that you can progress at your own pace.

You might be reading this book because you're considering going back to school and need to pass a college entrance exam or certification course. Maybe you're a professional working in finance, engineering, or management who finds that refreshing your math skills could give you an edge in your career. Or maybe you're a parent who wishes to support your child's education but feels rusty with the math concepts they are studying. No matter which category you fall into, this book provides the tools you need to regain your math confidence and competency.

Just picture being able to sit down with your child's homework and confidently explain how to solve a problem. Think about walking into a job interview or meeting with a fresh understanding of statistical methods that impresses your future employer or colleagues. Or imagine the satisfaction of acing that college entrance exam and taking the first step toward a new educational journey. These scenarios are within reach, and this book is your guide to getting there.

When you address common pain points and focus on real-world applications, you will make sure that you don't just learn math for the sake of passing a test, but understand how these concepts directly apply to your life and work. Whether it's calculating interest rates, budgeting for a project, or simply making sense of everyday statistics, the chapters are filled with relevant examples that make the material relatable and practical.

Additionally, to supplement your learning experience, we've included a **downloadable workbook** full of extra exercises, practice problems, and solutions to help you further solidify your understanding of each concept. This workbook is designed to complement the book's content, giving you more hands-on opportunities to apply what you've learned in a structured way.

To enhance a sense of community among readers, we've included stories and testimonials from people just like you—adults who once grappled with math but found their footing through structured relearning. These stories serve to inspire but also prove that it's never too late to improve your math skills. You're part of a diverse group of learners, each with unique experiences and goals, united by the shared objective of conquering math.

Going through this book is simple, which gives you the flexibility to customize your learning experience to suit your specific needs. If you're looking for a comprehensive review, start from the beginning and work your way through each chapter. For those pressed for time or focused on particular areas, the book allows you to jump straight to the sections most relevant to you. Additionally, we offer a diagnostic test to help identify which areas require more attention, guiding you toward a more targeted learning approach.

Each chapter begins with an introduction to the key concepts covered, followed by step-by-step examples, practice problems, and tips for avoiding common pitfalls. We've also included answers and detailed solution breakdowns to help you verify your understanding and track your progress. And remember, there's no rush; take your time to fully grasp each concept before moving on.

We believe that math can be a powerful tool rather than an obstacle. Once you reach the end of the book, you'll not only have refreshed your math skills but will also have gained a renewed sense of confidence and empowerment. Whether you're tackling math for personal growth, professional success, or to assist others in their learning journey, this resource is designed to support you every step of the way.

So let's get started. Open the next page, download your workbook, and together, we'll begin demystifying math, one concept at a time. With patience, perseverance, and the right guidance, you're capable of mastering the math skills you need to succeed. Welcome to your math refresher journey—let's make it an enjoyable and rewarding experience.

CHAPTER 1

Numbers and Operations

This chapter explores the fundamental concepts of numbers and operations, including natural numbers, whole numbers, and negative integers. It goes into rational and irrational numbers and their roles in the number line, as well as the order of operations, PEMDAS. Through practical examples and exercises, you will gain confidence in handling these fundamental concepts.

Natural, Whole, Integer, Rational, and Irrational Numbers

When mastering mathematics, understanding different types of numbers and their unique properties is fundamental. These classifications not only build a solid foundation for more complex mathematical concepts but also enhance practical problem-solving skills in everyday scenarios. To kick off our exploration, we will first dig into natural numbers.

Natural Numbers

Natural numbers, starting from 1 and extending infinitely, are the simplest form of numbers we encounter in daily life. Think about when you count physical objects like apples or books. Every time you tally items, you are using natural numbers. These numbers are the building blocks for various other number types and are crucial for foundational math skills. For instance, when you learn to add or multiply, you're typically working with natural numbers first. This familiarity helps ease the transition to understanding more complex operations and different types of numbers.

Whole Numbers

Next, let's extend this idea slightly to whole numbers. Whole numbers include all natural numbers plus zero. While zero might seem insignificant, it plays a critical role in arithmetic and practical applications. Just imagine trying to represent the absence of an item without zero; it would make accounting and basic record-keeping extremely cumbersome. Whole numbers are integral in real-world applications, such as counting objects in inventory management or calculating budgets where zero can represent a neutral balance. Understanding whole numbers lays the groundwork for when you will encounter negative numbers later. The inclusion of zero bridges a critical gap, making it easier to grasp more advanced mathematical concepts.

Integers

Moving forward, we arrive at integers. Integers encompass whole numbers and their negative counterparts. Introducing negative numbers might initially seem daunting, but they are vital for understanding many concepts beyond basic arithmetic. Consider temperature readings: Negative values help us understand temperatures below freezing. In finance, dealing with debt requires a good grasp of negative numbers. Knowing that -5 means owing $5, while +5 means having $5, is essential in personal and professional financial planning. Also, integers play an important role in algebra, where variables can represent both positive and negative values. This understanding is critical for solving equations and inequalities that appear frequently in higher-level math and various professional fields.

Rational and Irrational Numbers

Let's now talk about rational and irrational numbers. Rational numbers are those that can be expressed as fractions, which means one integer divided by another (where the divisor is not zero). Examples include 1/2, 3/4, and -5/6. Rational numbers allow flexibility in representing parts of a whole, whether it's splitting a pizza into equal slices or calculating interest rates, where percentages often convert neatly into fractions. Understanding these numbers provides a foundation for tackling problems involving ratios, proportions, and even some statistical measures.

Let's quickly look at a few examples of rational numbers. Say that a pizza is cut into 4 equal slices. If you eat 3 of those slices, you've eaten 3/4 of the pizza. Now, if we put this into more mathematical terms, 3/4 is a rational number because it's one integer (3) divided by another non-zero integer (4). As a decimal, it can be expressed as 0.75, and as a percentage, it's 75%, but don't worry, we will get to these.

On the other hand, irrational numbers cannot be expressed as simple fractions. These numbers include famous constants like π (pi) and √2 (the square root of 2). They are known for their nonrepeating and non-terminating decimal expansions. Understanding irrational numbers is crucial for comprehending measurements and calculations that don't fit neatly into fractional forms. For example, in geometry, the value of π is essential for calculating the circumference and area of circles. Similarly, engineers and scientists frequently encounter irrational numbers when dealing with continuous data or natural phenomena where exact fractions aren't applicable. Also, we will talk more about these later; for now, it's only important to understand their differences.

Knowing the distinction between rational and irrational numbers enhances your grasp of the number line. On this line, rational numbers create neat intervals we can easily calculate and manipulate, while irrational numbers fill in the gaps, ensuring there are no "missing" points. This complete understanding supports various applications, such as interpreting statistical data where precise values are necessary or scientific contexts where approximations are frequent yet crucial.

Order of Operations (PEMDAS)

PEMDAS is your mathematical compass. Standing for Parentheses, Exponents, Multiplication, Division, Addition, and Subtraction, this device outlines the critical sequence for tackling complex calculations. Adhering to this order is non-negotiable; deviating from it can lead to wildly inaccurate results.

The PEMDAS Hierarchy Explained:

1. Parentheses (P): Begin with operations enclosed in parentheses. These groupings demand immediate attention, as they are the foundation for the entire expression. For instance, in 7 x (4 + 3), you'd first compute 4 + 3 = 7, transforming the expression to 7 x 7 = 49. Parentheses provide clarity and structure to your calculations.

2. Exponents (E): Next, we tackle any exponents present. These shorthand notations represent repeated multiplication. In the expression 3^2 + 5, calculate 3^2 first: 3 x 3 = 9, resulting in 9 + 5 = 14. Exponents are particularly crucial in scientific and financial contexts. Consider (5 + 1)^3 - 2^2. First, simplify within parentheses: (6)^3 - 2^2. Then, evaluate exponents: 216 - 4 = 212

3. Multiplication and Division (MD): Proceed with multiplication and division operations, working from left to right. This left-to-right progression is vital for accuracy. Examine 24 ÷ 4 x 3. First, 24 ÷ 4 = 6, then 6 x 3 = 18. Reversing this order would yield an incorrect result.

4. Addition and Subtraction (AS): Finally, addition and subtraction are performed, again moving from left to right. In the expression 15 - 7 + 2,

start by subtracting 7 from 15 to get 8, then add 2 for a final answer of 10. Consistency in direction is key to avoiding errors in these final steps.

Practical Examples

To cement your understanding, let's explore some practical examples where the order of operations plays a critical role. Imagine you're handling your monthly budget and need to calculate the total expenses. Suppose you have $200 for groceries, $150 for utilities, and a one-time purchase of $50 spread over three months.

Using PEMDAS:

Step 1: Divide 50 by 3, which gives approximately 16.67.

Step 2: Add the amounts: 200 + 150 + 16.67.

Thus, the total is $366.67.

Such real-life scenarios underline the importance of mastering PEMDAS to avoid financial missteps.

Common Mistakes and Misconceptions

Common mistakes often arise when learners rush through problems without paying attention to the rules. One frequent error is neglecting to address parentheses first. Consider the expression 5 × (2 + 3) versus 5 × 2 + 3. The former equals 5 × 5, giving you 25, while the latter gives 10 + 3, equating to 13. The difference between the two results highlights why strict adherence to PEMDAS is essential.

Another common pitfall is mishandling multiplication and division. Remember, these operations are carried out from left to right, regardless of their position in the equation. Misinterpreting this can lead to incorrect outcomes and frustration. A helpful tip is to write out each step, ensuring that no part of the equation is skipped or overlooked.

Critical thinking and double-checking are vital habits to cultivate. Solving complex expressions accurately requires careful scrutiny to avoid simple errors. Always revisit each step, reconfirm the sequence, and ensure clarity in your calculations. Over time, this methodical approach will become second nature, bolstering your confidence in tackling math problems independently.

Exercises and Applications

To further enhance your learning, engage with various exercises and applications. Practice problems ranging from simple to complex will solidify your grasp of the order of operations.

Let's solve the following expression step-by-step: $(8 + 2)$ x $3^2 \div (1 + 1) - 5$.

1. Parentheses: Start by solving the operations inside the parentheses: $(8 + 2) = 10$ $(1 + 1) = 2$. Now, the expression is: 10 x $3^2 \div 2 - 5$

2. Exponents: Next, calculate the exponent: $3^2 = 9$ The expression now becomes: 10 x $9/2 - 5$

3. Multiplication and Division: Perform multiplication and division from left to right: First, 10 x $9 = 90$. Then, $90/2 = 45$; So, the expression is now: $45 - 5$

4. Addition and Subtraction: Finally, handle addition and subtraction from left to right: $45 - 5 = 40$

The final answer is 40.

Incorporate similar problems into your daily practice to gain proficiency. Real-life applications, such as adjusting recipes or balancing checkbooks, offer meaningful and practical ways to apply these principles. When you consistently practice this, you'll build a solid foundation and confidently tackle more advanced mathematical concepts.

Diagnostic Quizzes on Numbers and Operations

Okay, so let's practice. You will find the solutions for these expressions at the end of the book. But don't peek before you try to solve them; this is the only way you will learn.

1. $10 + 2 \times 3$

2. $(8 + 4)/2 - 3$

3. $5^2 + 3 \times 4 - 7$

As we've seen, this chapter goes into numbers, including natural, whole, integers, rational, and irrational numbers. Mastering these types prepares you for advanced math concepts and makes abstract ideas more relevant. Understanding these numbers helps tackle complex tasks and solve practical problems efficiently. The next chapter expands on this foundation, focusing on fractions and decimals.

CHAPTER 2

Fractions and Decimals

This chapter explores the basics of fractions and decimals, focusing on how to convert them. It covers their functions, their usage in different contexts, and their conversion techniques. Practical applications, such as financial calculations and everyday measurements, show the importance of these skills in avoiding common mistakes.

Converting Between Fractions and Decimals

Mastering the ability to convert between fractions and decimals is a crucial skill for anyone looking to refresh their math knowledge.

Understanding the Basics

Before diving into conversion techniques, it's important to understand what fractions and decimals are and their uses in various contexts. Fractions represent parts of a whole and are written as one number over another, separated by a slash (e.g., 1/2). The numerator, written above the line, represents the quantity of parts we're considering, whereas the denominator, written below the line, specifies the total number of equal divisions in the whole. Fractions are often used to describe quantities that are less than one but can also represent numbers greater than one in the form of improper fractions (e.g., 5/4).

In contrast, decimal notation employs a period (decimal point) to distinguish between the integer portion and the fractional component of a number. Each position after the decimal point represents a power of ten (e.g., 0.5, 0.25). Decimals are widely used in everyday life, particularly in financial transactions, measurements, and statistical data. Understanding both forms helps you interpret and manipulate numerical information more effectively.

Conversion Techniques

Now that we have a basic understanding let's explore how to convert fractions to decimals and vice versa. Learning these step-by-step techniques will help make this process straightforward.

Converting Fractions to Decimals

Transforming a fraction into its decimal equivalent involves performing division: the numerator is divided by the denominator. For instance, to express 3/4 as a decimal, calculate 3 divided by 4, which yields 0.75. Here are some steps to follow:

1. **Divide the numerator by the denominator:** Use long division if needed.

2. **Place the decimal point correctly:** If the division doesn't come out evenly, continue dividing until you reach a repeating pattern or a satisfactory decimal place.

For instance, 5/8 converts to 0.625 because 5 divided by 8 equals 0.625.

Converting Decimals to Fractions

To convert a decimal to a fraction, follow these steps:

1. **Count the decimal places:** Determine the number of digits to the right of the decimal point.

2. **Write as a fraction:** Place the decimal number over the appropriate power of 10. For example, 0.75 becomes 75/100.

3. **Simplify the fraction:** Reduce the fraction to its simplest form by finding the greatest common divisor (GCD) and dividing both the numerator and the denominator by it.

Using our example, for 0.75:

- Count two decimal places → 75/100

- Simplify by dividing both the numerator and the denominator by 25 (the GCD of 75 and 100) → {75/25} / {100/25} = 3/4

Common Mistakes

Despite the relative simplicity of these conversions, some common mistakes can occur. Identifying and correcting these errors will build confidence and accuracy.

One frequent error is misplacing the decimal point during conversion. For example, confusing 0.25 for 2.5 can lead to significant calculation errors. Always double-check the placement of the decimal point and ensure it matches the intended value.

Another mistake is forgetting to simplify fractions. When converting a decimal like 0.50 to a fraction, you get 50/100, which simplifies to 1/2. Skipping the simplification step can result in unnecessarily complex fractions that are harder to work with.

Incorrectly interpreting repeating decimals is also a common issue. For instance, 1/3 converts to 0.333... (repeating). It's vital to understand and properly notate repeating decimals to avoid confusion in further calculations.

Arithmetic Operations With Fractions

Performing arithmetic operations with fractions and mixed numbers can seem daunting at first, but once you grasp the basic principles, it becomes much more manageable. This section will guide you through addition, subtraction, multiplication, and division of fractions and mixed numbers, offering practical techniques and guidelines along the way.

Adding and Subtracting Fractions

Let's start with addition and subtraction of fractions. The key to adding or subtracting fractions is to have a common denominator. As we've seen, a denominator is the bottom number of a fraction, and a common denominator is necessary because it essentially represents the same "whole," which allows you to combine the parts accurately.

Let's solve the following addition step-by-step: 1/4 + 1/6

1. Find the Least Common Multiple (LCM): Start by finding the LCM of the denominators: LCM of 4 and 6 = 12. Now, we'll use 12 as our common denominator.

2. Convert Fractions: Next, convert each fraction to an equivalent fraction with the common denominator: 1/4 = (1 x 3)÷(4 x 3) = 3/12; 1/6 = (1 x 2)÷(6 x 2) = 2/12 The expression now becomes: 3/12 + 2/12

3. Add Numerators: Perform addition of the numerators, keeping the common denominator: (3 + 2)÷12 = 5/12

4. Simplify (if possible): Check if the resulting fraction can be simplified: 5/12 is already in its simplest form.

The final answer is 5/12.

But let's take a step back and figure out how we can find the LCM. Let's take the example above: 4 and 6. First, we list the multiples of each step. So, the multiples of 4 are: 4, 8, 12, 16, 20, 24, and so on. The multiples of 6 are: 6, 12, 18, 24, etc. As mentioned above, you need to find the smallest number that appears on both lists (the LCM), which in this case would be 12.

For subtraction, the process is quite similar. Suppose you want to subtract 1/8 from 1/3. Again, you need a common denominator.

Let's solve the following subtraction step-by-step: 1/3 - 1/8

1. Find the LCM: Start by finding the LCM of the denominators: LCM of 3 and 8 = 24. Now, we'll use 24 as our common denominator.

2. Convert Fractions: Next, convert each fraction to an equivalent fraction with the common denominator: 1/3 = (1 x 8)÷(3 x 8) = 8/24; 1/8 = (1 x 3)÷(8 x 3) = 3/24. The expression now becomes: 8/24 - 3/24

3. Subtract Numerators: Perform subtraction of the numerators, keeping the common denominator: (8 - 3)÷24 = 5/24

4. Simplify (if possible): Check if the resulting fraction can be simplified: 5/24 is already in its simplest form.

The final answer is 5/24.

So, just to recap:

1. Find a common denominator.

2. Convert each fraction to an equivalent fraction with the common denominator.

3. Add or subtract the numerators while keeping the denominator the same.

4. Simplify the resulting fraction, if possible.

Multiplication and Division of Fractions

Next up is multiplication and division of fractions. Multiplying fractions is straightforward. You just multiply the numerators together and the denominators together. For instance, multiplying 2/3 by 4/5 involves multiplying 2 by 4 and 3 by

5, resulting in 8/15. No need for a common denominator here, making multiplication often simpler than addition or subtraction.

Division involves flipping the second fraction (the divisor) and then multiplying. This flipped version is called the reciprocal. For example, to divide 3/4 by 2/5, you flip 2/5 to get its reciprocal, which is 5/2, and then you multiply: 3/4 x 5/2 = 15/8.

So, to recap:

1. For multiplication, multiply the numerators together and the denominators together.

2. For division, flip the second fraction (the divisor) to get its reciprocal and multiply.

3. Simplify the resulting fraction, if possible.

Working With Mixed Numbers

Now, let's move on to working with mixed numbers, which are whole numbers combined with fractions, like 2 1/2. To perform arithmetic operations with mixed numbers, it's often easiest to convert them into improper fractions first. An improper fraction has a numerator larger than its denominator. For example, converting 2 1/2 involves multiplying the whole number by the denominator (2 x 2) and adding the numerator, resulting in 5/2.

When adding or subtracting mixed numbers, convert them to improper fractions, follow the appropriate rules, and then convert back to a mixed number if needed.

Let's solve the following addition step-by-step: 3 1/4 + 2 2/3

1. Convert Mixed Numbers to Improper Fractions: Start by converting each mixed number to an improper fraction: 3 1/4 = (3 x 4 + 1) ÷ 4 = 13/4; 2 2/3 = (2 x 3 + 2) ÷ 3 = 8/3 The expression now becomes: 13/4 + 8/3

2. Find the LCM of the denominators: LCM of 4 and 3 = 12. Now, we'll use 12 as our common denominator.

3. Convert Fractions: Convert each fraction to an equivalent fraction with the common denominator: 13/4 = (13 x 3)÷(4 x 3) = 39/12; 8/3 = (8 x 4)÷(3 x 4) = 32/12 The expression now becomes: 39/12 + 32/12

4. Add Numerators: Perform addition of the numerators, keeping the common denominator: (39 + 32)÷12 = 71/12

5. Convert Back to Mixed Number: Convert the improper fraction to a mixed number: 71 ÷ 12 = 5 remainder 11 So, 71/12 = 5 11/12

The final answer is 5 11/12.

Multiplying and dividing mixed numbers follows a similar pattern. Let's solve the following multiplication step-by-step: 1 3/5 x 2 1/4

1. Convert Mixed Numbers to Improper Fractions: Start by converting each mixed number to an improper fraction: 1 3/5 = (1 x 5 + 3) ÷ 5 = 8/5; 2 1/4 = (2 x 4 + 1) ÷ 4 = 9/4 The expression now becomes: 8/5 x 9/4

2. Multiply Numerators and Denominators: Multiply the numerators together and the denominators together: (8 x 9) ÷ (5 x 4) = 72/20

3. Simplify the Fraction: Simplify the resulting fraction if possible: 72/20 = 18/5 (both numerator and denominator are divisible by 4)

4. Convert Back to Mixed Number: Convert the improper fraction to a mixed number: 18 ÷ 5 = 3 remainder 3 So, 18/5 = 3 3/5

The final answer is 3 3/5.

1. Convert mixed numbers to improper fractions before performing any operations.

2. Follow the addition, subtraction, multiplication, or division rules for improper fractions.

3. Convert your result back to a mixed number, if required.

4. Simplify where possible.

Diagnostic Quizzes on Fractions and Decimals

Finally, to consolidate this knowledge, let's tackle some challenges and practice problems. Practice is essential for mastering fractions and mixed numbers. Here are some problems you can work through:

1. Add the following fractions: 5/6 + 7/8.

2. Subtract 3/5 from 7/10.

3. Multiply 2/3 by 3/7.

4. Divide 9/11 by 3/4.

5. Add the mixed numbers: 4 2/3 and 1 5/6.

These problems are designed to test your knowledge and make you comfortable with the concepts discussed. Don't rush through them; take your time and ensure you understand each step before moving on. Converting between fractions and decimals might seem tricky at first, but with patience and practice, it gets easier. Here, we explained the basics of fractions and decimals, their conversion, and practical applications. The next chapter will focus on percentages and ratios, further deepening understanding.

CHAPTER 3

Percentages and Ratios

In this chapter, we will look at the fundamental skills of calculating percentages and ratios, essential for personal and professional purposes. We will also talk about percentage increases and decreases as well as its formulas. Lastly, we will talk about ratios.

Percentage Increase and Decrease Calculations

Calculating percentage changes is an essential skill for making informed financial and business decisions. A percentage change measures how a value has either increased or decreased relative to its original amount. This subpoint will break down the process of calculating percentage increases and decreases, illustrate these concepts with examples, and provide practice problems to cement your understanding.

Understanding Percentage Increase

Percentage increase is used to understand how much a value has grown over time. This can be crucial for assessing performance improvements or growth rates in various contexts, such as investments, sales, or production levels. The formula for calculating percentage increase is straightforward:

Percentage Increase = [(New Value - Original Value) ÷ Original Value] × 100

The new value is the value after the increase, and the original value is the initial value before the increase. So, first, you have to subtract the original value from the new value. Then, you have to divide the result by the original value, and lastly, you have to multiply by 100 to convert to a percentage.

Let's calculate the percentage increase in a company's revenue from last year to this year:

- Last year's revenue: $50,000

- This year's revenue: $60,000

  1. Calculate the Increase: Start by finding the difference between the new value and the original value: Increase = New Value - Original Value Increase = $60,000 - $50,000 = $10,000 The revenue increased by $10,000.

  2. Divide the Increase by the Original Value: Calculate the ratio of the increase to the original value: Ratio = Increase ÷ Original Value Ratio = $10,000 ÷ $50,000 = 0.2

  3. Convert to Percentage: Multiply the ratio by 100 to get the percentage: Percentage Increase = Ratio x 100 Percentage Increase = 0.2 x 100 = 20%

The final answer is 20%.

Understanding this calculation helps you gauge the extent of growth accurately, which is essential for setting realistic goals and expectations.

Common Pitfalls in Calculating Percentage Increases

While calculating percentage increases, there are common pitfalls to be aware of. One typical error is confusing the base value. Always use the original value as the base when calculating the increase. Using the new value instead will result in incorrect calculations.

For instance, if someone mistakenly uses the new revenue ($60,000) as the base in our previous example:

[($60,000 - $50,000) ÷ $60,000] x 100

($10,000 ÷ $60,000) x 100

0.1667 x 100

16.67%

This miscalculation shows a lower growth rate than the actual 20%, leading to underestimations that could affect decision-making.

Understanding Percentage Decrease

Next, let's discuss percentage decrease. This measurement helps determine how much a value has declined, which is important for budgeting and expense tracking. The formula for calculating percentage decrease is:

Percentage Decrease = [(Original Value - New Value) ÷ Original Value] × 100

Here, the original value is the initial value before the decrease, and the new value is the value after the decrease. To calculate it, you first need to subtract the new value from the original value. Then, divide the result by the original value and multiply by 100 to convert to a percentage.

Let's calculate the percentage decrease in a company's expenses:

- Original expenses: $30,000

- New expenses: $25,000

1. Calculate the Decrease: Start by finding the difference between the original value and the new value: Decrease = Original Value - New Value Decrease = $30,000 - $25,000 = $5,000 The expenses decreased by $5,000.

2. Divide the Decrease by the Original Value: Calculate the ratio of the decrease to the original value: Ratio = Decrease ÷ Original Value Ratio = $5,000 ÷ $30,000 = 0.1667 (rounded to 4 decimal places)

3. Convert to Percentage: Multiply the ratio by 100 to get the percentage: Percentage Decrease = Ratio x 100; Percentage Decrease = 0.1667 x 100 = 16.67%

The final answer is 16.67%.

Alerts on Miscalculating Reductions

As with percentage increases, there are common errors to watch out for when calculating percentage decreases. A frequent mistake is failing to subtract correctly or using the wrong values for the original and new amounts.

If you use the new value ($25,000) as the base instead of the original value, you'd have this:

[($30,000 - $25,000) ÷ $25,000] × 100

($5,000 ÷ $25,000) × 100

0.2 × 100

20%

This error shows a reduction of 20% instead of 16.67%, potentially leading to misleading conclusions about expense management.

Combined Examples for Better Retention

To solidify these concepts, let's examine combined examples. Suppose a small business sees their monthly profits rise from $5,000 to $6,500 and then fall back to $5,800 the following month. We'll calculate both the percentage increase and decrease.

First, the percentage increase from $5,000 to $6,500:

[($6,500 - $5,000) ÷ $5,000] × 100

($1,500 ÷ $5,000) × 100

0.3 × 100

30%

The business saw a 30% increase in profits between months.

Next, the percentage decreases from $6,500 to $5,800:

[($6,500 - $5,800) ÷ $6,500] × 100

($700 ÷ $6,500) × 100

0.1077 × 100

10.77%

The business experienced approximately a 10.77% decrease in profits between months.

These examples illustrate how the same data set can be analyzed from different angles, providing a comprehensive view of performance trends.

Practice Problems

Finally, it's time to put your knowledge to the test with practice problems. Working through these exercises will help reinforce your skills and build confidence in calculating percentage changes.

1. A store's inventory costs increased from $15,000 to $18,000. Calculate the percentage increase.

2. A company's employee count decreased from 120 to 95. Determine the percentage decrease.

3. Sales revenue grew from $45,000 to $52,000, then dropped to $49,000. Calculate both the percentage increase and decrease.

Solutions:

Problem 1

Given:

A. Original Value: $15,000

B. Calculate the Increase: Increase = New Value - Original Value Increase = $18,000 - $15,000 = $3,000

The inventory costs increased by 20%.

A. Calculate the Percentage Increase: Percentage Increase = (Increase ÷ Original Value) x 100 Percentage Increase = ($3,000 ÷ $15,000) x 100 Percentage Increase = 0.2 x 100 = 20%

B. New Value: $18,000

Problem 2

Given:

A. Original Value: 120 employees

B. New Value: 95 employees

C. Calculate the Decrease: Decrease = Original Value - New Value Decrease = 120 - 95 = 25

D. Calculate the Percentage Decrease: Percentage Decrease = (Decrease ÷ Original Value) x 100 Percentage Decrease = (25 ÷ 120) x 100 Percentage Decrease = 0.2083 x 100 = 20.83%

The employee count decreased by approximately 20.83%.

When you practice these problems and discuss your findings with peers or family, you can enhance collaborative learning and ensure a solid grasp of percentage changes.

Ratio Problem-Solving Techniques

Understanding ratios is essential for interpreting numeric relationships and making well-informed decisions in various real-life scenarios. Ratios provide a way to compare two quantities, simplifying complex problems into manageable figures. Let's dive into the basics of ratios and their applications.

Basics of Ratios

A ratio represents the relationship between two quantities. For example, if a recipe calls for 2 cups of flour and 1 cup of sugar, the ratio of flour to sugar is 2:1. This means there are two parts flour for every one-part sugar. Ratios can be written in different forms, such as 2:1, 2/1, or "2 to 1."

One important aspect of working with ratios is simplifying them to their lowest terms for clarity. This involves dividing both parts of the ratio by their greatest common divisor (GCD). For instance, a ratio of 8:4 can be simplified to 2:1 by dividing both numbers by 4, the GCD. Simplifying ratios makes it easier to understand and work with the data.

Solving Ratio Problems

Solving ratio problems requires a methodical approach. Identify the known and unknown quantities and determine the relationship between them through the given ratio. Here are some steps to solve basic ratio problems effectively:

1. **Read and Understand**: Fully comprehend the problem by identifying the quantities involved and the ratio provided.

2. **Set Up the Ratio**: Write down the ratio in fraction form, ensuring you align the parts correctly.

3. **Scale the Ratio**: Determine if the ratio needs to be scaled up or down to match the given conditions of the problem.

4. **Solve for the Unknown**: Use algebraic methods to solve for the unknown quantity.

Consider an example where you have a mixture that consists of blue and red beads in a 3:5 ratio. If there are 15 red beads, how many blue beads are there? Set up the ratio like this: let the number of blue beads be x. The ratio equation will be $3/5 = x/15$. Cross-multiplying gives $3 \times 15 = 5x$, leading to $45 = 5x$, and solving for x gives $x = 9$. So, there are nine blue beads.

Let's break this down. The ratio 3:5 means that for every 3 blue beads, there are 4 red beads. This can be written as a fraction like this: 3/5 (blue/red). Let x = the number of blue beads (the unknown), and we know there are 15 red beads. The ratio of blue to red should be the same as 3:5. This means we can write: $3/5 = x/15$. Which can also read as "3 is to 5 as x is to 15."

1. Set Up the Proportion: Ratio of blue to red beads: 3:5. Number of red beads: 15. Let x be the number of blue beads. Set up the proportion: 3/5 = x/15

2. Cross-Multiply: Multiply the numerator of each fraction by the denominator of the other: (3 x 15) = (5.x) 45 = 5x

3. Solve for x: Divide both sides by 5 to isolate x: 45 ÷ 5 = 5x ÷ 5 9 = x

4. Check the Answer: Verify that 9:15 simplifies to 3:5 9 ÷ 3 = 3 15 ÷ 3 = 5 The ratio 9:15 does simplify to 3:5, confirming our answer.

The final answer is 9 blue beads.

Now we've found that x = 9 which means that there are 9 blue beads, but we can verify if this fits into the original ratio. So, we assume that [blue : red] = [9 : 15]. We can simplify this ratio: 9/3 : 15/3 = 3 : 5. This matches our original ratio, confirming our answer is correct.

Guidelines for solving ratio problems can be beneficial:

- Always simplify the ratio first.

- Carefully read and re-read the problem to avoid misunderstandings.

- Use proportionate scaling to relate the given information to the ratio.

- Verify your solution by substituting back into the problem.

Diagnostic Quizzes on Percentage and Ratios

Challenge problems push readers to apply these concepts critically. Here are a few challenging problems to test your understanding:

1. A store is having a sale where all items are marked down by 15%. What is the sale price of an original $80 jacket?

2. In a school of 500 students, 60% play sports. Of those who play sports, 40% play basketball. How many students play basketball?

3. A school has students in the ratio of 7 boys to 5 girls. If there are 210 boys, how many girls are there?

Carefully solving these problems step-by-step will strengthen your grasp of percentages and ratios.

In this chapter we explored percentages, ratios, and proportions, focusing on their calculation and application in business performance, budgeting, and everyday situations. Besides that, we went into formulas and common pitfalls to allow accurate calculations and practical application in financial decisions and investments. In the next chapter, we will be looking at factors and multiples.

CHAPTER 4

Factors and Multiples

Determining factors and multiples is a fundamental skill in mathematics that sets the stage for solving more complex problems. Factors are numbers you can divide into another number without leaving a remainder, while multiples are what you get when you multiply a number by an integer. This chapter offers practical insights into calculating the LCM and GCD, which are critical tools in various mathematical applications, as we've already seen.

Prime Factorization Methods

Prime factorization is a key mathematical concept that helps in understanding the factors of numbers and plays an important role in solving various types of math problems. It involves breaking down a number into its prime components, which are the building blocks of all numbers. Prime numbers are those that are greater than one and have no divisors other than one and themselves. Examples include 2, 3, 5, 7, and 11. These numbers hold special significance because they cannot be divided further. Understanding prime numbers is essential to grasping the process of prime factorization.

Understanding Prime Numbers

When we talk about prime factorization, it means expressing a number as a product of prime numbers. This technique is useful for determining the factors of a given number efficiently. One common method for finding the prime factors of a number is the factor tree method. This visual tool helps break down the steps in a clear, organized manner, making it easier to understand.

Let's find the prime factorization of 60 using the factor tree method.

1. Start with 60

2. Divide by the smallest prime factor (2): 60 = 2 x 30

3. Continue with 30: 30 = 2 x 15

4. Factor 15: 15 = 3 x 5

5. The factor tree is complete when all factors are prime: 60 /
   2 30 /
   2 15 /
   3 5

6. Write the prime factorization: 60 = 2 x 2 x 3 x 5

7. Express using exponents: 60 = 2^2 x 3 x 5

The final prime factorization of 60 is 2^2 x 3 x 5.

Now, let's discuss a practical guideline for using the factor tree method:

1. Start with the smallest prime number (2) and divide the given number.

2. Divide by the same prime number until you can't divide evenly.

3. Move to the next smallest prime number and repeat the process.

4. Continue this process until the only remaining numbers are prime numbers.

Understanding how to list all factors of a number using its prime factorization is another important aspect. Once a number is expressed as a product of primes, we can use these primes to determine all possible factors. For instance, to find all the factors of 60, we use its prime factorization (2^2 x 3 x 5). Listing all combinations of these prime factors gives us: 1, 2, 3, 4 (which is (2^2)), 5, 6 (which is (2 x 3)), 10 (which is (2 x 5)), 12 (which is (2^2 x 3)), 15 (which is (3 x 5)), 20 (which is (2^2 x 5)), 30 (which is (2 x 3 x 5)), and 60.

A good guideline for listing all factors using prime factorization is as follows:

1. Start with the prime factorization of the number.

2. List all possible products of these prime factors.

3. Include 1 and the number itself.

Using prime factorization to determine the GCD is a practical application. The GCD of two or more numbers is the largest number, and it divides all of them without leaving a remainder. To find the GCD, we compare the prime factorizations of the numbers involved and identify the common prime factors.

Let's take two numbers as an example: 48 and 180. The prime factorization of 48 is (2^4 x 3), and for 180, it is (2^2 x 3^2 x 5). The common prime factors are 2 and 3. We then take the lowest power of these common factors present in both factorizations. For 2, the lowest power is (2^2), and for 3, it is (3^1). Thus, the GCD is (2^2 x 3 = 4 x 3 = 12).

Here's a guideline for finding the GCD using prime factorization:

1. Find the prime factorization of each number.

2. Identify the common prime factors.

3. Take the lowest power of each common prime factor.

4. Multiply these lowest powers together to get the GCD.

Least Common Multiple Calculations

The LCM plays a crucial role in various mathematical applications. To start, let's define what it is and why it's important. The LCM of two or more integers is the smallest positive integer that is divisible by all the given numbers without leaving a remainder. We've talked about this, but let's refresh your memory. Consider the numbers 4 and 5. The multiples of 4 are 4, 8, 12, 16, 20, and so on. The multiples of 5 are 5, 10, 15, 20, and so forth. The smallest multiple they have in common is 20. Thus, 20 is the least common multiple of 4 and 5.

Now, let's get into how to calculate the LCM through prime factorization. Prime factorization is the process of decomposing a number into a product of its prime constituents. For example, the prime factors of 12 are $2^2 * 3$, while the prime factors of 18 are 2×3^2. To find the LCM using these prime factorizations, take the highest power of each prime number involved. In this case, the highest power of 2 is 2^2, and the highest power of 3 is 3^2. Thus, the LCM is $2^2 * 3^2$, which equals 36.

Let's go through this step-by-step:

1. **Factor Each Number**: Begin by listing the prime factors for each number.

   - For 12: $2^2 \times 3$

   - For 18: 2×3^2

2. **Identify Highest Powers**: Identify the highest power of each prime number present in any of the factorizations.

   - 2: The highest power is 2^2.

- 3: The highest power is 3^2.

3. **Multiply These Values**: Multiply these highest powers together to get the LCM.

   - LCM = $2^2 \times 3^2 = 4 \times 9 = 36$

This method ensures that you capture all the prime factors required without missing any shared factors.

Alternative Methods for LCM Calculation

However, prime factorization isn't the only way to determine the LCM. Another method is listing out the multiples of each number until you find the smallest number they share. Take 8 and 12, for example. The multiples of 8 are 8, 16, 24, 32, etc., and the multiples of 12 are 12, 24, 36, etc. The smallest common multiple here is 24, so the LCM of 8 and 12 is 24.

Another alternative approach is using division. This method works well if you are dealing with larger numbers. Here's how:

1. **Write Down the Numbers**: List the numbers for which you are finding the LCM.

2. **Divide by Common Prime Factors**: Divide these numbers by their common prime factors until all you have left are prime numbers.

3. **Continue Until All Numbers Are Reduced to 1**: Keep dividing by any common prime numbers until there are no more common factors left other than 1.

4. **Multiply All Divisors Together**: The product of these divisors gives you the LCM.

For example, let's find the LCM of 14 and 20 using division:

1. 14 and 20

2. Both can be divided into 2:

    - $14 \div 2 = 7$

    - $20 \div 2 = 10$

3. Now we have 7 and 10. There are no common prime factors between them.

4. Multiply the original divisor (2) by the remaining numbers 7 and 10:

    - $LCM = 2 * 7 * 10 = 140$

These methods ensure that you can find the LCM efficiently depending on the numbers and context you're working with.

Diagnostic Quizzes on Factors

Find the LCM using prime factorization.

1. Find the LCM of 18 and 24

2. Calculate the LCM of 15, 25, and 30

3. Determine the LCM of 56 and 72

Reflection Questions on Arithmetic Foundations

Okay, so now let's revise your arithmetic foundations by solving some exercises on the material we've covered so far.

Numbers and Operations

1. Evaluate the following expression: $12 + 3 \times (8 - 2^2) \div 2$

2. Simplify: $72/9 + 5 \times 4 - 12$

3. Calculate: $(36 - 12 \times 2) + (8 + 16/4)$

Fractions and Decimals

1. Add the following fractions: 3/8 + 5/12

2. Multiply: 2/3 x 3/4

3. Convert 0.625 to a fraction in its simplest form

Percentage and Ratios

1. In a class of 40 students, 15 are boys. What percentage of the class are girls?

2. A shirt priced at $50 is on sale for 20% off. What is the sale price?

3. The ratio of cats to dogs in a pet store is 5:3. If there are 24 cats, how many dogs are there?

Factors and Multiples

Find the LCM of the following sets of numbers using prime factorization:

1. 18 and 24

2. 36, 48, and 60

3. 42 and 70

This chapter discussed prime factorization and its importance in simplifying math problems. We used the factor tree method to illustrate each step and provide practical guidelines. Understanding that prime factorization helps in finding the GCD, it simplifies calculating the LCM. In the next chapter, we are going to change the subject (well, sort of) by looking at the foundations of algebraic concepts.

CHAPTER 5

Basic Algebraic Concepts

Without a grasp of variables, expressions, and the techniques used to simplify them, advancing to more complex topics can be daunting. This chapter is designed to ease learners into these fundamental ideas, making the journey less overwhelming and more intuitive. The two main parts of algebraic expressions that we will examine in this chapter are variables and constants. We'll look into how these elements interact through operators like addition, subtraction, multiplication, and division. We'll also discuss methods for simplifying algebraic terms by combining like terms, providing visual aids, and using hands-on practice problems to reinforce learning. Through clear explanations and practical examples, you will gain a robust understanding of these basic yet crucial algebraic concepts, setting the stage for mastering more advanced mathematical topics in future chapters.

Simplifying Algebraic Expressions

An algebraic expression consists of variables and constants combined using mathematical operators such as addition, subtraction, multiplication, or division. Grasping how these elements come together will provide clarity on the structure of algebra and offer a solid foundation for future learning.

Symbols, typically alphabetic characters that denote undetermined or mutable quantities are called variables. In contrast, constants are unalterable values that remain fixed throughout calculations or operations. When these two elements are joined with operators, they form expressions. For example, in the expression (3x + 5), (x) is a variable, while 3 and 5 are constants. Understanding this distinction is crucial because it sets the stage for manipulating and simplifying algebraic expressions.

The Role of Operators

Operators like addition, subtraction, multiplication, and division play a critical role in shaping the value of these expressions. Each operator has a specific function and order of operations that must be followed to arrive at the correct result. For instance, consider the expression (2x + 4 - 3x). Here, both the addition and subtraction operators are used. The order in which we perform these operations affects the outcome of the expression.

To demystify this further, let's focus on the impact of each operator. Addition and subtraction combine terms to increase or decrease the total value. In contrast, multiplication scales the value of the variable, and division reduces it. When we break down expressions step by step and observe the results, we can better understand how each operator influences the overall value.

Example of Simplifying Expressions

Consider an example where we simplify the expression (3x + 4 - 2x + 9). First, we identify like terms—those with the same variable component. In this case, (3x) and (-2x) are like terms, as are 4 and 9. Combining the like terms, we get:

[3x - 2x = x] [4 + 9 = 13]

Thus, the simplified expression becomes (x + 13). Visualizing each step through diagrams or drawings can help solidify this process in the reader's mind.

Practice Problems

Hands-on practice problems are vital for cementing these skills. You can do this by engaging directly with examples, applying what you've learned, and gaining confidence in your abilities. Let's look at a few practice problems to illustrate this point:

1. Simplify the expression: (5x + 3 - 2x + 7) Step-by-step:

   - Identify like terms: (5x) and (-2x), 3 and 7.

   - Combine the like terms: (5x - 2x = 3x).

   - Add the constants: (3 + 7 = 10).

   - Final result: (3x + 10).

Working through these problems allows you to experience firsthand how to manipulate and simplify algebraic expressions. This active participation not only reinforces theoretical knowledge but also builds problem-solving skills critical for more advanced topics.

Recognizing and Combining Like Terms

In algebra, understanding and working with like terms is a foundational skill that significantly eases the simplification of expressions. Identifying like terms involves recognizing common features among different algebraic terms. Essentially, like terms are terms that contain the same variables raised to the same powers. For instance, in the expression (3x + 5x - 2y), both (3x) and (5x) are like terms since they both feature the variable (x), making them combinable. This identification can be a stepping stone toward grasping more complex algebraic concepts by revealing predictable patterns within equations.

Figuring out how to identify like terms can make those seemingly complicated algebraic structures much more familiar and less intimidating. Consistency is key:

If learners remember that like terms share the same variable parts, this consistency paves the way for clearer and more structured thought processes. It's like sorting apples from oranges; once you know your categories, the task becomes straightforward, allowing you to get into algebraic operations with confidence.

Visual representations can be extremely beneficial in further elucidating the process. Diagrams, charts, and other forms of illustrations simplify the idea by breaking down each step visually. When tackling $(3a + 4a - 2b)$, drawing out the grouping of like terms can make it more comprehensible: $(3a)$ and $(4a)$ pair together while $(-2b)$ stands alone. These visuals act as guides, providing learners with a reference point that translates abstract numbers and variables into something more concrete and understandable.

Example Problems

Once you have grasped the concept with the help of visuals, practicing through exercises becomes invaluable. Providing immediate feedback during these practice sessions is crucial for reinforcing their newfound skills. For example, an exercise could involve simplifying the expression $(6m + 2n - 4m + n)$.

1. Like terms with m: 6m and -4m; like terms with n: 2n and n

2. Group-like terms: $(6m - 4m) + (2n + n)$

3. Combine like terms: $6m - 4m = 2m$; $2n + n = 3n$

4. Write the simplified expression: $2m + 3n$

The final simplified expression is $2m + 3n$.

As learners work through such exercises, discussing the solutions immediately afterward makes sure that any misunderstandings are promptly addressed. Feedback should highlight both the correct steps taken and any mistakes made, offering constructive insights on how to avoid similar errors in the future. This not only helps in retaining the correct methods but also builds self-assurance.

Diagnostic Quizzes on Basic Algebra Concepts

1. Simplify: $3x + 2y - 5x + 4y - 7$

2. Simplify $2(x + 3) - 3(x-1)$

3. Identify the like terms in the expression: $5x^2 + 3xy - 2x^2 + 4y - xy + 2$

Gaining a strong understanding of fundamental algebraic concepts like variables, constants, and operators is important. Practice makes perfect, as you know, and mastering each tool individually will prepare you for more complex tasks.

CHAPTER 6

Working With Integers and Exponents

Working with integers and exponents is a fundamental skill in mathematics that builds the foundation we've been working on until now when it comes to tackling more complex problems. This chapter goes through the intricacies of integer arithmetic and the rules governing exponents, making them accessible for anyone eager to refresh their mathematical knowledge.

Integer Arithmetic Operations

Understanding arithmetic operations involving integers is key to building strong calculation skills and improving confidence in handling mathematical problems. This section will clarify the rules for addition, subtraction, multiplication, and division of integers, along with mixed operations.

Addition and Subtraction of Integers

Let's begin with adding and subtracting positive and negative integers. Visualization techniques, such as using a number line, can greatly help in comprehending how these operations work. Picture a number line stretched horizontally with zero in the center, positive numbers extending rightward, and negative numbers extending

---

leftward. Advancing rightward along the number line corresponds to the addition of a positive whole number. Conversely, when adding a negative integer, move to the left.

For example, consider the equation 5 + (-3).

1. Start at 5 on the number line.

2. Adding a negative number means moving left: Move 3 units to the left

3. Identify the final position: You land on 2

Therefore, 5 + (-3) = 2

For -4 + 6 = 2:

1. Start at -4 on the number line.

2. Adding a positive number means moving right: • Move 6 units to the right

3. Identify the final position: • You land on 2

Therefore, -4 + 6 = 2

Subtracting integers follows similar rules but with a twist. Subtracting a positive integer means moving left while subtracting a negative integer involves moving right, effectively converting the operation into addition.

Example 1: 7 - 3

1. Start at 7 on the number line.

2. Subtracting a positive number means moving left: Move 3 units to the left

3. Identify the final position: You land on 4

Therefore, 7 - 3 = 4

Example 2: -2 - (-5)

1. Start at -2 on the number line.

2. Subtracting a negative number is the same as adding it's positive: Convert -2 - (-5) to -2 + 5

3. Adding a positive number means moving right: Move 5 units to the right

4. Identify the final position: • You land on 3

Therefore, -2 - (-5) = 3

Multiplication of Integers

Let's move on to the rules governing integer multiplication. The product of two integers depends on their signs. Multiplying two positive integers or two negative integers yields a positive result. Multiplying one positive and one negative integer results in a negative outcome. These sign rules form the backbone of integer multiplication.

Let's examine three examples of multiplying integers with different sign combinations.

Example 1: 4 x 3 (Positive x Positive)

1. Identify the signs: Both integers are positive.

2. Multiply the absolute values: 4 x 3 = 12

3. Determine the sign of the product: Positive x Positive = Positive

4. Result: 4 x 3 = 12

Example 2: -4 × -3 (Negative x Negative)

1. Identify the signs: Both integers are negative.

2. Multiply the absolute values: 4 x 3 = 12

3. Determine the sign of the product: Negative x Negative = Positive

4. Result: -4 x -3 = 12

Example 3: 5 × -2 (Positive x Negative)

1. Identify the signs: One positive, one negative.

2. Multiply the absolute values: 5 x 2 = 10

3. Determine the sign of the product: Positive x Negative = Negative

4. Result: 5 x -2 = -10

Practical applications highlight these rules. Imagine calculating the profit or loss in financial trades. A positive gain times the number of successful trades remain positive. However, a loss per trade (negative value) times the number of trades would result in an overall loss (negative value).

Division of Integers

Integer division shares similarities with multiplication in terms of sign rules. Dividing two positive integers or two negative integers gives a positive quotient. Dividing one positive and one negative integer results in a negative quotient. It's essential to understand these principles to avoid common pitfalls in calculations.

Example 1: 20 ÷ 4 (Positive ÷ Positive)

1. Identify the signs: Both integers are positive.

2. Divide the absolute values: 20 ÷ 4 = 5

3. Determine the sign of the quotient: Positive ÷ Positive = Positive

4. Result: 20 ÷ 4 = 5

Example 2: -20 ÷ -4 (Negative ÷ Negative)

1. Identify the signs: Both integers are negative.

2. Divide the absolute values: 20 ÷ 4 = 5

3. Determine the sign of the quotient: Negative ÷ Negative = Positive

4. Result: -20 / -4 = 5

However, mistakes often occur during integer division. A common error is neglecting the sign rule, which can lead to incorrect quotients. To prevent this, always verify the signs before concluding any division operation.

Combined Operations

Transitioning to mixed operations involving integers requires careful attention to detail and adherence to the order of operations, often remembered by the acronym PEMDAS (just to refresh your memory, this stands for: Parentheses, Exponents, Multiplication and Division, Addition and Subtraction). When faced with equations combining multiple operations, perform calculations within parentheses first, followed by exponents (if any), then tackle multiplication and division from left to right, and finally, address addition and subtraction from left to right.

Let's solve the expression: -3 + 6 x (2 - 8)/2

1. Parentheses: Solve operations within parentheses first (2 - 8) = -6 The expression becomes: -3 + 6 x -6/2

2. Exponents: There are no exponents in this expression

3. Multiplication and Division (left to right): a. 6 x -6 = -36 The expression becomes: -3 + -36/2 b. -36/2 = -18 The expression becomes: -3 + -18

4. Addition and Subtraction (left to right): -3 + -18 = -21

The final answer is -21.

Working through these types of problems repeatedly helps solidify understanding and ensures that the transition between different types of operations becomes second nature.

Exponentiation Rules

As we've seen, an exponent represents how many times a number, known as the base, is multiplied by itself. For example, in the expression (2^3), 2 is the base and 3 is the exponent. This notation indicates that 2 should be multiplied by itself three times, resulting in $(2 \times 2 \times = 8)$. Visualizing exponents can help solidify this concept. Imagine starting with a single unit (1) and doubling it repeatedly; the result displays the power of exponential growth clearly.

Product of Powers Rule

When dealing with exponents, certain rules simplify calculations significantly. A core principle in exponent operations is the Law of Exponents for Multiplication, also known as the Product of Powers Rule. This rule states that when you multiply two expressions with the same base, you add their exponents.

Rule: $a^m \times a^n = a^{(m+n)}$

Example: Simplify $2^3 \times 2^4$

1. Identify the base and exponents: Base: 2; First exponent (m): 3; Second exponent (n): 4

2. Apply the rule: $2^3 \times 2^4 = 2^{(3+4)} = 2^7$

3. Calculate the result: $2^7 = 128$

Therefore, $2^3 \times 2^4 = 128$

Think about calculating the volume of a cube with side length equivalent to a base raised to different powers. Multiplying these volumes aligns perfectly with adding the exponents.

Power of Powers Rule

Next, let's discuss the Power of a Power Rule. This rule is applied when an exponent is raised to another exponent, which simplifies to multiplying the exponents together.

Rule: (a^m)^n = a^(m×n)

Example: Simplify (3^2)^4

1. Identify the base and exponents: Base: 3; Inner exponent (m): 2; Outer exponent (n): 4

2. Apply the rule: (3^2)^4 = 3^(2x4) = 3^8

3. Calculate the result: 3^8 = 6561

Therefore, (3^2)^4 = 6561

Intuitive examples make this clear. Imagine folding a piece of paper in half (doubling its thickness) and then folding it in half again repeatedly. The number of layers increases exponentially, illustrating how quickly powers can escalate.

Negative Exponents and Zero Exponents

Understanding negative exponents can initially seem tricky, but they're quite logical once broken down. An exponent with a minus sign denotes the inverse of the base elevated to the equivalent positive power.

Rule: a^(-n) = 1 ÷ a^n

Example: Simplify 2^(-3)

1. Identify the base and exponent: Base: 2; Exponent: -3

2. Apply the rule: $2^{(-3)} = 1 \div 2^3$

3. Calculate the result: $1 \div 2^3 = 1 \div 8 = 0.125$

Therefore, $2^{(-3)} = 1/8$ or 0.125

This relationship highlights why dividing by the base repeatedly results in smaller and smaller fractions. Zero exponents are equally straightforward; any non-zero base raised to the power of zero equals one. This is expressed as $(a^0 = 1)$, regardless of the base. It's a reflection of the notion that multiplying a number by itself zero times leaves you with a multiplicative identity, which is always one.

To apply these principles effectively, here's a guideline for the Product of Powers Rule and Power of a Power Rule:

Product of Powers Rule:

- Identify if the bases in multiplication are the same.

- Add the exponents while keeping the base unchanged.

- Simplify the resulting expression accordingly.

Power of a Power Rule:

- Notice if an exponent is raised to another exponent.

- Multiply the exponents while retaining the base.

- Evaluate the final expression.

Being able to use these rules accurately allows for a seamless transition into more complex algebraic manipulations since these foundational concepts recur throughout various fields of math. For instance, professionals in engineering or finance frequently simplify large expressions involving repeated multiplications, and these rules streamline their calculations.

Diagnostic Quizzes on Integers and Exponents

1. Evaluate: $(-3)^4$

2. Simplify: $2^5 \times 3^2$

3. Simplify: $(x^2)^3$

Ultimately, mastering exponentiation rules like the Product of Powers and Power of a Power provides a sturdy foundation for diving deeper into mathematics. These rules are not only pivotal for academic success but also for practical problem-solving in real-life scenarios. Embracing these concepts with clarity and confidence prepares learners for both higher education and the numerous professional challenges they might face.

In this chapter, we looked at the basics of exponents and integer operations, preparing you for complex mathematical problems. We emphasized the importance of understanding these concepts for mastering other mathematical areas, enhancing computational abilities, and showing the changing power of a strong mathematical foundation in daily life and career.

CHAPTER 7

Understanding Proportions and Rates

Understanding proportions and rates will help you simplify many aspects of your daily life. Whether it's comparing prices while shopping, adjusting recipes, or planning a trip, being able to set up and solve proportions helps make informed decisions with ease. This chapter will guide you through the practical applications of these concepts, showing how they can be seamlessly integrated into everyday situations. Here, you'll also learn how to identify and solve proportions, starting with simple problem statements and moving on to more complex word problems. The techniques covered include cross-multiplication and strategies for setting up proportional equations from real-world scenarios. We'll also look at how you can calculate unit rates, which allows for comparing different quantities like price per item or miles per hour.

Proportion Problem Strategies

Understanding proportions is a critical skill, especially for adults returning to education or professionals seeking to refresh their math skills. Proportions are essentially equations that express the equality of two ratios. Knowing how to identify and solve them can make many real-life tasks—such as adjusting recipes, calculating distances, or comparing prices—much simpler.

Identifying Proportions

One of the foundational skills in working with proportions is identifying them within problem statements. This involves recognizing pairs of quantities that maintain a consistent relationship. For instance, if you read a statement like "3 apples cost $6; how much do 5 apples cost?" The question sets up a proportion between apples and their cost. You notice that the ratio of apples to dollars (3:6) should be equivalent to the unknown ratio (5:x), where x represents the cost of 5 apples. This simple identification helps set the stage for solving the problem more effectively.

Cross-Multiplication Technique

Once you've identified a proportion, the next step is solving it, and one reliable method is cross-multiplication. Cross-multiplication is an easy-to-apply technique that simplifies the process of solving proportions. Using the previous example (3/6 = 5/x):

Let's solve the proportion: 3/6 = 5/x

1. Set up the proportion: 3/6 = 5/x, where x is the unknown cost for 5 apples

2. Cross-multiply: Multiply the numerator of each fraction by the denominator of the other (3.x) = (6 x 5)

3. Simplify the right side: 3x = 30

4. Solve for x: Divide both sides by 3 3x/3 = 30/3 x = 10

5. Check the solution: Substitute x = 10 into the original proportion 3/6 = 5/10 1/2 = 1/2 (This checks out)

6. Interpret the result: 5 apples would cost $10

Key points to remember:

- Cross-multiplication is a technique for solving proportions.

- To verify your result, invariably insert the derived solution back into the initial proportional equation.

- Interpret the result in the context of the problem.

Guidelines for cross-multiplication involve several critical steps. First, ensure that both ratios are written in fraction form. Then, cross-multiply by taking the numerator of one fraction and multiplying it by the denominator of the other, performing this operation for both fractions. Finally, solve the resulting equation for the unknown variable.

Developing Word Problems

Another important aspect of working with proportions is developing word problems and translating them into proportional equations. Word problems often describe real-world scenarios where proportions play a vital role. When restating the problem as a proportion, you can streamline the process of finding a solution. Let's say you're helping your child with homework and come across a question about mixing paint. The problem states that to create a certain shade, you need 2 parts blue paint to 3 parts yellow paint. If you need 15 parts of yellow paint, how much blue paint do you need? Setting up the proportion ($2/3 = x/15$), you quickly cross-multiply to find that x equals 10 parts of blue paint.

Hands-On Examples

Crafting word problems also hones your ability to think critically about everyday situations. Consider another scenario, such as budgeting expenses. Suppose you spend $200 on groceries for four weeks. How much would you expect to spend over six weeks? Here, the proportion is $200/4 = x/6$. Cross-multiplying (as we did above) gives you $4x = 1200$, and solving for x tells you that you would spend $300 over six weeks. Such exercises help you see the practical applications of proportions in managing finances and other daily tasks.

To further enhance your understanding, it's beneficial to include worked-through examples. These examples serve as a bridge between theory and practice, showcasing various strategies to solve proportions. Let's take a look at a detailed example. Imagine you're planning a road trip and trying to figure out fuel costs. Your vehicle gets 25 miles per gallon, and the last time you checked, gas was $3 per gallon. How much would it cost to drive 150 miles? First, determine the number of gallons needed by using the proportion 25/1 = 150/x, which results in x equaling 6 gallons. Next, multiply the number of gallons by the price per gallon to find that 6 gallons at $3 each cost $18. But let's break it down one more time:

The given information is:

- Vehicle efficiency: 25 miles per gallon

- Gas price: $3 per gallon

- Trip distance: 150 miles

First, we set up the proportion to find the number of gallons needed. Here, we can use the ratio of miles to gallons:

25 miles/1 gallon = 150 miles/x gallons. We can write the equation this way: 25/1 = 150/x.

Then, solve the proportion using cross-multiplication:

Cross-multiply: 25x = 150 x 1, which is simplified as 25x = 150.

Then, solve for x (number of gallons) by dividing both sides by 25. x = 150/25.

This means that x = 6 gallons. Then, multiply the number of gallons by the price per gallon:

6 gallons x $3 per gallon = $18. So, the fuel cost for the 150-mile trip would be $18.

Another example might involve cooking. Say you have a recipe that serves 4 people, but you want to adjust it to serve 10.

If the recipe calls for 2 cups of flour for 4 servings, how many cups are needed for 10 servings?

Set up the proportion 2/4 = x/10. Cross-multiplying yields 4x = 20, so x equals 5 cups of flour.

Diagnostic Quizzes on Proportions and Rates

1. The ingredients list specifies a ratio of 2 units of sugar to 5 units of flour in the culinary instructions. If you want to use 8 cups of flour, how many cups of sugar do you need?

2. In a class, the ratio of boys to girls is 3:4. If there are 28 girls, how many boys are in the class?

3. A car travels 210 miles in 3.5 hours. At this rate, how far will it travel in 5 hours?

Reflection Questions on Pre-Algebra Essentials

Let's now revise what we've learned so far on pre-algebraic essentials.

Basic Algebraic Concepts

1. Simplify: $3x + 2y - 5x + 4y - 7$

2. Combine like terms: $2a^2 - 3ab + 5a^2 + 2ab - 4a^2$

3. Simplify: $4(x + 2) - 2(x - 3)$

Integers and Exponents

1. Evaluate: $(-7) + 12 - (-5) + (-3)$

2. Calculate: $(-4) \times 3 \times (-2)$

3. Simplify: 15 - 3 x (-2) + 4/(-2)

Proportions and Rates

1. The culinary instructions say a proportion of 3 parts flour to 2 parts sugar in the ingredient composition. If you want to use 5 cups of sugar, how many cups of flour do you need?

2. A car travels 240 miles in 4 hours. At this rate, how far will it travel in 6 hours?

3. If 15 pencils cost $3.75, what is the cost of 24 pencils?

In this chapter, we've explored how to set up and solve proportions, making it easier to handle everyday tasks like adjusting recipes, calculating travel costs, or comparing prices. Identifying proportions in problem statements is key, and once you do that, using methods like cross-multiplication becomes straightforward. We've also looked into creating word problems and translating them into proportional equations, offering practical examples to show their real-world applications. We also explored unit rates and their importance in our lives. In the next chapter, we are once again changing the tune and exploring the fundamentals of geometry.

CHAPTER 8

Geometry Fundamentals

Geometry is all about understanding the shapes, sizes, and properties of different figures. This chapter dives into the essential building blocks of geometry, focusing on points, lines, and angles. Each of these components plays a crucial role in forming more complex geometric concepts. It doesn't matter if you're designing a new building or simply trying to understand the layout of your living room; a solid grasp of these basics will set you up for success.

Here, we'll look at how to identify and differentiate various types of angles—acute, right, obtuse, and straight—and see how they appear in everyday contexts. We'll also cover practical techniques for measuring angles and line segments accurately using tools like protractors and rulers. Once we're done, you'll be equipped with skills like estimating angles quickly and calculating perimeters and areas of different shapes. These foundational skills will not only help you in academic settings but also have real-world applications in fields ranging from engineering to interior design.

Different Types of Angles

We'll look at the foundational knowledge needed to identify and differentiate between various types of angles. Recognizing angles in real-world contexts is an essential skill, enhancing one's ability to apply geometric concepts in everyday life and professional scenarios.

Acute Angles

First up, let's talk about acute angles. Acute angles measure less than 90 degrees. These are the small, sharp corners you might see almost everywhere, from the pointed ends of a star design to the sleek lines in architecture. When architects plan buildings, they often use acute angles to add visual interest and modern appeal to structures. Consider the slanted roofs on houses or the triangular shapes in bridges; these angles help distribute weight efficiently while also adding aesthetic value. An acute angle can give a structure a sense of dynamism and movement, making it not just functional but also visually appealing.

Right Angles

Now, shifting our focus to right angles, which measure exactly 90 degrees. Right angles are incredibly common and super important in everyday life. Just look around your room—door frames, books, computer screens—almost all of these objects have right angles. They're fundamental in construction because they ensure things are stable and square. Imagine building a bookshelf; if the angles weren't right angles, the shelf wouldn't hold items securely, leading to imbalance and potential collapse. In technical fields like engineering and carpentry, right angles are the cornerstone for creating functional, reliable designs. They provide the perfect balance and symmetry required for both simple tasks and complex projects.

Obtuse Angles

Moving on to obtuse angles, which measure more than 90 degrees but less than 180 degrees. These angles often show up in geometric designs and artistic applications. Think about a sprawling fan shape or the design of a large slice of pizza—both illustrations of obtuse angles. In the world of interior design, obtuse angles can create spacious and open environments. For example, designing rooms with walls that meet at obtuse angles can make spaces appear larger and more inviting. In graphic design, using obtuse angles helps to break the monotony of straight lines, adding a bit of flair and creativity to layouts. Such angles are instrumental in crafting visually compelling and functional spaces or designs.

Straight Angles

Lastly, we have straight angles that measure exactly 180 degrees. When two rays form a straight line, you've got yourself a straight angle. These angles are crucial in construction and civil engineering. For instance, roads and railways often feature straight angles to ensure they follow a direct path, maximizing efficiency and minimizing travel time. The concept of a straight angle also extends to furniture design, where pieces like benches and tables use straight lines to promote simplicity and functionality. In structural designs, straight angles help distribute force evenly, ensuring stability and strength. Whether it's planning a new roadway or arranging furniture in a living space, straight angles play a pivotal role in creating coherent, efficient designs.

Measurement Techniques in Geometry

Accurately measuring angles, lines, and geometric shapes is a basic skill in geometry. This section goes into the essential measurement techniques needed to master these basic components.

Using a Protractor

A protractor is an indispensable tool for anyone dealing with angles. Whether you're a student preparing for exams or a professional working on a design project, knowing how to use a protractor ensures precision in your measurements. To measure an angle with a protractor, place the midpoint of the protractor at the vertex of the angle. Align one side of the angle with the zero line on the protractor. The number where the other side intersects the protractor's scale indicates the measure of the angle. Practice using the protractor on different angles until you feel comfortable with the process. Over time, you'll find that this tool becomes an extension of your hand.

However, you might not remember exactly what the zero line on the protractor is. As we've mentioned above, this is the starting point for measuring angles. The zero line is often found at the base of the protractor, aligning the straight edge. Most protractors have two zero lines, one at each end of the straight edge. This allows for measuring angles in both clockwise and counterclockwise directions. Another thing to take into consideration is that when you are measuring an angle, one ray of the angle should be aligned with the zero line.

Measuring Line Segments

When it comes to measuring line segments, precision is equally vital. Tools like rulers and tape measures are commonly used for this purpose. A ruler is best for small distances, while a tape measure is handy for longer lengths. To measure a line segment with a ruler, place the ruler along the segment so that the zero on the ruler aligns with one end of the segment. Then, read the number on the ruler where the other end of the segment meets the scale. Keep the ruler steady and ensure it's straight to get an accurate measurement. Practice measuring various objects around you, from books to tables, to build your confidence.

Tape measures are often used in fields like construction and interior design, where larger distances need to be measured. Using a tape measure follows similar principles to using a ruler but requires careful handling to avoid inaccuracies caused by bending or slack in the tape. Hold the tape measure firmly, pull it taut, and make sure it remains straight for an accurate reading.

Estimating Angles

While tools like protractors provide precise measurements, there are situations where estimating angles can be incredibly useful. Quick estimation skills are particularly beneficial in professions requiring swift decision-making. To estimate an angle, start by visualizing common angles you are familiar with, such as 90 degrees for a right angle or 180 degrees for a straight line. Compare the angle you need to estimate with these reference points. With enough practice, your accuracy in estimating angles will improve.

Engage in activities that encourage angle estimation, such as sketching simple geometric shapes without using a protractor or observing and guessing the angles formed by items in your environment. For instance, look at the corner of a book or a window frame and try to estimate the angle. Check your estimate with a protractor to see how close you were. This practice sharpens your sense of angles and makes you more adept over time.

Calculating Perimeters and Areas

Knowing how to calculate perimeters and areas is crucial for many real-world applications, from designing a garden layout to planning a room renovation. The perimeter refers to the distance around a shape, while the area measures the space within it.

To calculate the perimeter of a rectangle, add the lengths of all four sides together. For example, if a rectangle has two sides of 5 meters and two sides of 3 meters, its perimeter would be (5 + 3) x 2 = 16 meters. For regular polygons like squares, where all sides are equal, multiply the length of one side by the total number of sides. So, if you have a square and its side is 8 meters, then you have to do 8 x 4 = 32 meters. Let's break this down:

Calculating the area involves multiplying the base by the height for rectangles and squares. The aforementioned rectangle's area would be 5 meters x 3 meters = 15 square meters. Different shapes have different formulas for area calculation. For example, the area of a triangle is calculated by multiplying the base by the height and then dividing by two. So, for instance, if a triangle has a base of 8 centimeters and a height of 6 centimeters first, you have to recall its formula: Area = (1/2) x base x height. We know that the base (b) = 8 cm and the height (h) = 6 cm. So, if we use the formula, we have: Area = (1/2) x 8 cm x 6 cm. This means that Area = (1/2) x 48 cm^2; therefore, Area = 24 cm^2 (note that the exponent in this case means square meters; you don't have to solve it).

Start with simple shapes and standard formulas before moving on to more complex figures. Practicing with different shapes and forms helps solidify your understanding and prepares you for practical applications. With this said, let's look at some basic formulas to calculate the area:

- Square: Area = s^2(where s is the length of a side)

- Rectangle: Area = l × w(where l is length and w is width)

- Triangle: Area = (1/2) × b × h(where b is base and h is height)

- Circle: Area = π r^2(where r is radius and π is approximately 3.14159)

- Parallelogram: Area = b × h(where b is base and h is height)

- Trapezoid: Area = (1/2)(a + b)h(where a and b are the parallel sides and h is the height)

Diagnostic Questions on Geometric Fundamentals

1. A rectangle has a length of 12 cm, and a width of 8 cm. Calculate its perimeter and area.

2. A square garden has a side length of 15 meters. Find its perimeter and area.

3. A circular pizza has a diameter of 14 inches. Calculate its circumference and area. (Use π = 3.14)

This chapter has provided a thorough introduction to the basics of geometry, helping readers understand essential components like points, lines, and angles. These fundamental concepts are crucial for grasping more advanced geometric ideas and have practical applications across various fields. While exploring different types of angles—acute, right, obtuse, and straight—we've seen how these shapes are present in our daily lives and professional activities. Whether it's the design of buildings using acute angles or ensuring stability with right angles, understanding these basics enhances one's ability to apply geometry in real-world situations.

CHAPTER 9

Shapes and Their Properties

Shapes are everywhere in our world, from the rectangles of our mobile phones to the circles of our dinner plates. Understanding shapes and their properties can make a huge difference in both everyday tasks and professional projects. Whether you're calculating the amount of paint needed for a wall or planning the layout of a garden, knowing how to find the perimeter, area, and volume of various shapes is invaluable. This chapter delves into the nitty-gritty details of 2D and 3D shapes, guiding you through essential concepts and practical calculations.

You'll start by exploring the perimeter and area of 2D shapes like squares, rectangles, triangles, and circles. Learn the specific formulas that make these calculations straightforward and see how they apply in real-world scenarios. As you gain confidence with these foundational skills, you'll then transition into the fascinating world of 3D shapes. From cubes and rectangular prisms to cylinders and spheres, this chapter will teach you how to determine the volume of these objects. By the end, you'll not only be proficient in crunching numbers but also prepared to tackle complex shapes and avoid common pitfalls in your calculations. Whether you're brushing up on math for professional reasons, helping your child with homework, or preparing to return to education, this chapter has got you covered.

Perimeter and Area of 2D Shapes

Let's review what we've learned in the previous chapter before moving on. Calculating the perimeter and area of various 2D shapes is an essential skill that can be applied in numerous real-life situations.

First, let's delve into how to measure the distance around a shape using specific formulas. The perimeter is the total length of all sides of a 2D shape. For squares and rectangles, this calculation is straightforward. To find the perimeter of a square, simply multiply the length of one side by four ($P = 4a$). For a rectangle, add together the lengths of all four sides or use the formula $P = 2(l + w)$, where 'l' stands for length and 'w' stands for width. Applying this to a practical example: If you're planning to fence a yard that's rectangular and measures 50 feet by 30 feet, you'd calculate the perimeter as $2(50 + 30) = 160$ feet, ensuring you know how much fencing material you need.

1. Identify the formula: Perimeter of rectangle = 2(length + width)

2. Substitute the values: Perimeter = 2(50 ft + 30 ft)

3. Calculate: Perimeter = 2(80 ft) = 160 ft

Therefore, you need 160 feet of fencing material.

Triangles, with their three sides, have a slightly different approach. To find the perimeter, sum the lengths of all three sides ($P = a + b + c$). For instance, if a triangular garden has sides measuring 10 feet, 7 feet, and 5 feet, its perimeter would be $10 + 7 + 5 = 22$ feet. This knowledge is particularly useful when outlining garden beds or constructing triangle plots.

1. Identify the formula: Perimeter of triangle = sum of all sides

2. Add the lengths of all sides: Perimeter = 10 ft + 7 ft + 5 ft

3. Calculate: Perimeter = 22 ft

Therefore, the perimeter of the triangular garden is 22 feet.

Circles require a different method involving pi (π). The perimeter of a circle, referred to as the circumference, is calculated using the formula C = $2\pi r$, where 'r' is the radius. If you wished to edge a circular flowerbed with a radius of 6 feet, you'd calculate the circumference as $2\pi \times 6$, which approximates about 37.68 feet.

1. Identify the formula: Circumference of circle = $2\pi r$, where r is the radius

2. Substitute the values: Circumference = $2\pi \times 6$ ft (Use $\pi \approx 3.14$ for approximation)

3. Calculate: Circumference = $2 \times 3.14 \times 6$ ft $\approx$ 37.68 ft

Therefore, you need approximately 37.68 feet of edging material.

Now, let's move on to calculating area, which refers to the space contained within a shape. For rectangles, use the formula A = l $\times$ w. If you're looking at laying down new sod on a rectangular lawn that measures 50 feet by 30 feet, the area calculation is straightforward: 50 $\times$ 30 = 1500 square feet.

1. Identify the formula: Area of rectangle = length $\times$ width

2. Substitute the values: Area = 50 ft $\times$ 30 ft

3. Calculate: Area = 1,500 ft²

Therefore, you need 1,500 square feet of sod.

When dealing with triangles, the area is determined using A = 1/2(b $\times$ h), where 'b' is the base and 'h' is the height. Imagine a triangular plot of land with a base of 8 feet and a height of 5 feet; the area would be 1/2(8 $\times$ 5) = 20 square feet. This calculation can come in handy for projects involving irregularly shaped plots or materials.

Circles follow the formula $A = \pi r^2$. If you were carpeting a circular room with a radius of 10 feet, the area would be $\pi \times 10^2$, approximating about 314.16 square feet. Understanding these basic calculations is crucial for home improvements, gardening, and various other practical applications.

Sometimes, you might encounter composite shapes—shapes made up of simpler geometric figures. To find the perimeter and area of these complex shapes, break them down into their constituent parts. Consider an L-shaped room composed of two rectangles. First, calculate the perimeter and area of each rectangle separately, then add these values appropriately. For instance, if one-part measures 15 feet by 10 feet and the other 12 feet by 8 feet, calculate each separately, then combine. The areas might be 150 square feet and 96 square feet, respectively, giving a total area of 246 square feet.

However, mistakes are common when computing perimeter and area. You may forget to include one side of a shape, miscalculate dimensions, or mix up formulas. To avoid these errors, always double-check your work. For instance, remeasure dimensions and ensure correct application of formulas. Logical reasoning helps too; if a calculation leads to an impractically large or small number, it's worth reviewing your steps.

When verifying perimeter and area, cross-referencing different methods can enhance accuracy. For example, if a rectangular room's perimeter seems off, recheck both the direct sum of all sides and the formula $2(l + w)$. Another technique is reconstructing composite shapes from scratch if initial results seem dubious, ensuring no dimensions are overlooked or misrecorded.

Volume Calculations for 3D Shapes

Now that we've recapped perimeters and areas of 2D shapes let's move on to 3D shapes. Understanding how to calculate the volume of different 3D shapes is important for various practical applications in everyday life. The quantity of space enclosed within a solid figure is known as its volume. Think of it as how much water you could fill inside a container or the capacity of a room. This concept is measured in cubic units, which represent a 1x1x1 unit cube fitting within the shape. Let's explore how to calculate the volume of some common 3D shapes.

To start, consider the simplest 3D shape: a cube. A cube's volume is found by cubing the length of one of its sides. The formula is ($V = a^3$), where (a) represents the side length. For instance, if each side of the cube is 2 meters long, the volume would be ($2 \times 2 \times 2 = 8$) cubic meters. This calculation helps determine how much space is available inside a box of that size.

- Identify the formula: Volume of cube = a^3, where a is the side length

- Substitute the value: $V = 2^3 \ m^3$

- Calculate: $V = 2 \times 2 \times 2 = 8 \ m^3$

Therefore, the volume of the cube is 8 cubic meters.

Next, examine the rectangular prism, another basic 3D shape commonly encountered in packaging and storage. The volume formula for a rectangular prism is ($V = l \times w \times h$), where (l) is the length, (w) is the width, and (h) is the height. Imagine a rectangular storage container with dimensions of 3 meters in length, 2 meters in width, and 4 meters in height. The volume would be ($3 \times 2 \times 4 = 24$) cubic meters. This formula helps calculate the capacity of various containers, from small boxes to large shipping crates.

1. Identify the formula: Volume of rectangular prism = $l \times w \times h$

2. Substitute the values: V = 3 m x 2 m x 4 m

3. Calculate: V = 24 m^3

The volume of the rectangular prism is 24 cubic meters.

Cylinders are frequently used for items like soup cans, water tanks, and pipes. To find the volume of a cylinder, use the formula ($V = \pi r^2 h$), where (r) is the radius of the base and (h) is the height. Suppose you have a cylindrical water tank with a radius of 1 meter and a height of 5 meters. The volume would be ($V = \pi \times 1^2 \times 5 = 5\pi = 15.7$ (approximately)) cubic meters. Understanding this formula is essential when dealing with round containers or columns in construction.

1. Identify the formula: Volume of cylinder = $\pi r^2 h$

2. Substitute the values: V = $\pi \times 1^2$ m^2 x 5 m (Use $\pi \approx 3.14$ for approximation)

3. Calculate: V = 3.14 x 1 x 5 = 15.7 m^3 (approximately)

So, the volume of the cylinder is approximately 15.7 cubic meters.

Spheres might seem less common, but they appear in objects like balls or decorative items. The formula for the volume of a sphere is ($V = \{4\} \div \{3\} \pi r^3$). If you have a basketball with a radius of 0.3 meters, the volume would be ($V = \{4\} \div \{3\} \pi \times 0.3^3 = 0.113$ (approximately)) cubic meters. Computing the volume of spheres helps in determining material requirements for manufacturing spherical objects or even calculating air capacity in balloons.

1. Identify the formula: Volume of sphere = $(4 \div 3)\pi r^3$

2. Substitute the value: V = $(4 \div 3) \times \pi \times 0.3^3$ m³^3(Use $\pi \approx 3.14$ for approximation)

3. Calculate: V = $(4 \div 3) \times 3.14 \times 0.027 \approx 0.113$ m^3

The volume of the sphere is approximately 0.113 cubic meters.

Cones also pop up in various contexts, including ice cream cones and traffic cones. To calculate the volume of a cone, use ($V = \{1\} \div \{3\} \pi r^2 h$). For example, an ice cream cone with a radius of 0.5 meters and a height of 1 meter has a volume of ($V = \{1\} \div \{3\} \pi \times 0.5^2 \times 1 = 0.26$ (approximately)) cubic meters.

1. Identify the formula: Volume of cone $= (1 \div 3)\pi r^2 h$

2. Substitute the values: $V = (1 \div 3) \times \pi \times 0.5^2 \, m^2 \times 1 \, m$ (Use $\pi \approx 3.14$ for approximation)

3. Calculate: $V = (1 \div 3) \times 3.14 \times 0.25 \times 1 \approx 0.26 \, m^3$

This means that the volume of the cone is approximately 0.26 cubic meters.

Sometimes, you'll need to find the volume of more complex shapes that aren't perfect: cubes, prisms, cylinders, spheres, or cones. In these cases, you can often break the complex shape into simpler components, like we did with 2D shapes. For instance, if you need to find the volume of a structure combining a rectangular prism and a half-sphere on top, first calculate the volume of the prism and then the half-sphere separately. Add the volumes together for the total volume. This technique is particularly useful in construction projects where precise material amounts are necessary.

For example, imagine constructing a pool with a rectangular base and semi-cylindrical ends. First, find the volume of the rectangular part using the prism formula. Then, calculate the volume of the semi-cylindrical ends using the cylinder formula divided by two and add these volumes together. This approach is practical for determining the amount of concrete required for specific shapes.

While calculating volume, it's easy to make mistakes, especially with unit conversions or incorrect formulas. Common pitfalls include not squaring or cubing measurements correctly or mixing up units. To avoid such errors, always double-check your calculations and ensure you're using the correct formulas for the shapes involved. Utilize the same measurement units throughout the calculation process to maintain consistency.

Finally, it's beneficial to verify your results with logical reasoning. For example, estimate whether the calculated volume seems reasonable based on the object's dimensions. If you're calculating the volume of a large room and end up with an unusually high or low number, reconsider your steps. Checking twice can save time and resources, especially in professional settings.

Diagnostic Quizzes on Shapes and Their Properties

1. A cube has a side length of 5 cm. Calculate its volume.

2. A rectangular prism has a length of 8 cm, a width of 6 cm, and a height of 4 cm. Find its volume.

3. A cylindrical water tank has a radius of 3 meters and a height of 10 meters. Calculate its volume. (Use $\pi = 3.14$)

Reflection Questions on Geometry Basics

1. A rectangle has a length of 15 cm and a width of 8 cm. Calculate its perimeter and area.

2. A triangle has a base of 10 m and a height of 6 m. If its other two sides are 8 m each, find its perimeter and area.

3. A cube has an edge length of 4 inches. Calculate its volume.

Understanding the properties and calculations of 2D and 3D shapes equips you with practical skills that can be applied in various real-life scenarios. Here, we talked about the perimeter, area, and volume of different geometric shapes. Whether it's calculating how much fencing material you need, determining the amount of sod for a lawn, or figuring out the capacity of a container, these mathematical concepts are essential tools. In the next chapter, we are moving on to a different concept and will be working with algebraic expressions.

CHAPTER 10

Working With Algebraic Expressions

Working with algebraic expressions involves understanding how to combine terms and use the distributive property efficiently. These skills are essential for manipulating expressions, solving equations, and grasping more advanced algebra concepts. Algebraic expressions form the basis for many mathematical operations, making it critical to master these basic techniques.

In this chapter, we will look into how to combine like terms, apply the distributive property, and practice various techniques for working with algebraic expressions. We will also have a look at how to identify and group similar terms to simplify expressions, making them easier to work with. The last subject we will get into is the distributive property, which allows you to expand or condense expressions as needed.

Factoring Algebraic Expressions

Knowing how to factor algebraic expressions allows readers to rewrite expressions in simpler forms, making solving equations easier and providing a foundation for further algebraic concepts. Factoring is the process of breaking down an expression into its constituent parts or factors, which, when multiplied together, give back the original expression. This process is fundamental because it simplifies complex expressions and reveals their structure, making them easier to manage and solve.

Methods of Factoring

In the context of algebraic expressions, factors are numbers or expressions that can be multiplied together to yield another number or expression. For example, in the expression $(12x^2 + 6x)$, both terms share a common factor of 6 and (x). When you factor out this common element, we can rewrite the expression as $(6x(2x + 1))$. This simplification makes it much easier to work with the expression in subsequent calculations or when solving equations. Let's break this down before we move on.

The first step is to identify the common factor, so you have to look at both terms and find what they have in common. So, the $12x^2$ has factors of 12 and x^2. And $6x$ has factors of 6 and x. This means the common factor is 6 and x. This still might be a little confusing, so let's have a closer look:

1. The first term $12x^2$ can be broken down into 12 (which can be further factored as 2 x 2 x 3, and we have x^2.

2. The second term, 6x, can be broken down into 6 (which can be factored as 2 x 3, and we have x.

3. You can then compare the factors. As you can see, both terms have 6 as a factor because 12 is divisible by 6. And both terms have x as a factor since x^2 in the first term includes x. Therefore, the largest common factor of both terms is 6x.

4. Then, you have to factor out the common factor, and to do this, you have to write the common factor outside parenthesis, which will give us the 6x(...) expression above.

5. After that, you divide each term by the common factor. So, for the first term, you do $12x^2 \div (6x) = 2x$. For the second term, you do $6x \div (6x) = 1$.

6. Once you get that, you write the remaining factors inside the parenthesis, which looks like this: $6x(2x + 1)$.

7. Then, all you have to do is check your work, and for that, you can distribute the 6x to ensure you get the original expression: $6x(2x + 1) = 6x(2x) + 6x(1) = 12x^2 + 6x$.

Knowing how to find the common factors simplifies expressions. When you identify and extract the greatest common factor (GCF) from each term within an expression, you streamline your calculations and reduce the potential for errors. The GCF is the highest number or variable that evenly divides all terms of the expression. This technique isn't just limited to numerical coefficients; it applies equally to variables shared among terms.

Next, let's look into various factoring techniques. To illustrate, consider the expression $(18y^3 + 24y^2 - 30y)$.

Here, the GCF is 6y, as it is the largest factor that each term shares. Factoring out the GCF, the expression becomes $(6y(3y^2 + 4y - 5))$. This process significantly simplifies the expression and sets the stage for further manipulation or solving.

Another important technique involves understanding special products like the difference of squares. A difference of squares occurs in expressions of the form $(a^2 - b^2)$, which can be factored into $((a + b)(a - b))$. Understanding this pattern is invaluable because it allows for quick simplifications that are often necessary in algebraic problem-solving.

For example, the expression $(x^2 - 25)$ can be factored into $((x + 5)(x - 5))$. Let us talk a little more about the difference between squares when it comes to factoring techniques using the example above.

This factoring technique is used for expressions in the form of $a^2 - b^2$, where a and b are any algebraic expressions. With this technique, you can factor such expressions into the product of two binominals: $(a + b)(a - b)$.

So, the pattern is $a^2 - b^2 = (a + b)(a - b)$. This works because if we multiply $(a + b)(a - b)$, we get $a^2 - ab + ab - b^2 = a^2 - b^2$. The middle terms (ab and -ab) cancel out, leaving us with $a^2 - b^2$.

If we look at the example given, $x^2 - 25$, we can identify $a = x$ (because x^2 is the first term) and $b = 5$ (because $5^2 = 25$).

This means we can factor $x^2 - 25$ as $(x + 5)(x - 5)$.

Here's a tip: when it comes to recognizing the difference of squares, look for an expression with two perfect square terms being subtracted. Also, keep in mind that this only works for subtractions, not additions (for instance, $a^2 + b^2$ cannot be factored this way).

Factoring Quadratics

When we discuss factoring quadratics, we touch on one of the most prevalent areas where factoring skills are applied. Quadratic expressions typically take the form $(ax^2 + bx + c)$. To factor quadratics, you first need to find two binomials that multiply to the original quadratic expression. Consider the quadratic $(x^2 + 5x + 6)$. Factoring this, we look for two numbers that multiply by 6 (the constant term) and add up to 5 (the coefficient of the linear term). Factors of 6 are 2 and 3, which also add up to 5. Thus, the expression factors into $((x + 2)(x + 3))$. Let me give you a few more examples.

- $x^2 + 7x + 12$; we need factors of 12 that add up to 7. So, 3 and 4 work because $3 \times 4 = 12$ and $3 + 4 = 7$.

- With this, we know that the factored form looks like $(x + 3(x + 4)$.

- $x^2 - 6x + 8$; we need factors of 8 that add up to -6.

- Here, -2 and - 4 work because $(-2) \times (-4) = 8$ and $(-2) + (-4) = -6$ and the factored form is $(x - 2)(x - 4)$.

Factoring quadratics also helps us find the zeros, or roots, of quadratic functions. These are the values of (x) that make the equation equal to zero. Using our previous example $(x^2 + 5x + 6 = 0)$, we factored it into $((x + 2)(x + 3) = 0)$. Setting each factor equal to zero gives $(x = -2)$ and $(x = -3)$, which are the zeros of the function.

Simplifying Polynomials

In this section, we aim to provide you with the tools needed to simplify polynomial expressions. Simplifying these expressions is essential for solving equations and grasping more complex algebra topics.

First, it's crucial to clarify what constitutes a polynomial. A polynomial is a mathematical expression composed of variables, coefficients, and constants, combined using addition, subtraction, and multiplication—but never division by a variable. Each part of a polynomial is called a term.

For instance, in the polynomial $(3x^2 + 4x - 5)$, there are three terms:

- $(3x^2)$

- $(4x)$

- (-5)

The number multiplying the variable is known as the coefficient. In our example, the coefficients are 3 and 4. Understanding the degree of a polynomial is equally important; it signifies the highest power of the variable within the expression. Here, $(3x^2)$ has the highest power (2), making it a second-degree polynomial.

Combining Like Terms

Once you have identified the components of a polynomial, the next step is to combine like terms efficiently. As we've seen in previous chapters, terms are terms that have identical variable parts raised to the same power. For example, $(2x^2)$ and $(5x^2)$ are like terms, but $(2x^2)$ and $(3x)$ are not.

Combining like terms simplifies polynomials, making them easier to work with.

Take the expression $(3x^2 + 2x + 5x^2 + 4x)$. Grouping like terms gives you $((3x^2 + 5x^2) + (2x + 4x))$, simplifying further to $(8x^2 + 6x)$.

Efficiency is key here. Grouping all like terms together before performing any arithmetic can help avoid mistakes. Practicing with various examples, such as combining $(7y + 3y - 2y)$, helps build confidence. This practice consolidates understanding and demonstrates its utility in everyday scenarios.

Applying Distributive Property

Next, let's discuss the distributive property, a critical tool for manipulating algebraic expressions. The distributive property states that $(a(b + c) = ab + ac)$. This means you can distribute a multiplier across terms inside parentheses. So, $(3(x + 4))$ becomes $(3x + 12)$. This property is particularly useful when dealing with multi-term expressions, transforming complex problems into simpler, more manageable ones.

Consider a practical example involving finance. Let's say you have a budget represented by the expression $(2(50 + 30) + 10 \times 4)$.

Using the distributive property, this can be simplified to $(2 . 50 + 2 . 30 + 10 . 4)$, which simplifies further to $(100 + 60 + 40 = 200)$. This makes calculations more straightforward, enhancing your ability to make informed financial decisions.

Let's recap quickly how you can apply the distributive property. Its formula is: a(b + c) = ab + ac. And in the example above, we had 3(x + 4). First, you distribute 3 to both terms inside the parenthesis 3(x) + 3(4). Then, simplify it: 3x + 12.

In the financial example, the original expression was 2(50 + 30) + 10 x 4. If you apply the distributive property, you get 2(50 + 30)(2 x 50) + (2 x 30) + 10 x 4. Then, you simplify each part: 100 + 60 + 40 and add the terms 200.

Note: At times, we will be using "." in the place of "x" in multiplications to avoid confusion with the variable "x."

Diagnostic Quizzes on Algebraic Expressions

1. Find the GCF and factor it out: $18x^3y^2 + 24x^2y^3$

2. Recognize and factor the difference of squares: $16a^2 - 81b^2$

3. Factor the following quadratic expression: $x^2 + 11x + 28$

This chapter has focused on teaching you how to combine terms and use the distributive property effectively. Once you learn how to do it, you will be able to handle algebraic expressions with greater ease, making your calculations faster and more accurate. We've explored various methods of factoring, from identifying the GCF to recognizing patterns like the difference of squares. These strategies simplify complex expressions, making them more manageable.

As you practice these techniques, you'll find that simplifying polynomials becomes easier. The ability to break down and simplify expressions will not only boost your confidence but also make solving equations less daunting. Keep practicing, and soon you'll see just how powerful and versatile these algebraic methods can be in real-world applications. In the following chapter, we will expand on this subject by looking at how to solve linear equations and inequalities.

CHAPTER 11

Solving Linear Equations and Inequalities

Solving linear equations and inequalities is quite important and serves as a foundation for various mathematical concepts and real-world applications. Understanding how to work through problems involving linear equations and inequalities enhances problem-solving abilities. We will be looking at the different methods and strategies for solving linear equations and inequalities. You'll learn how to graph solutions on a number line to visualize the results better. Then, we will discuss practical applications such as budgeting, project planning, and other scenarios where these mathematical tools are invaluable.

Graphing Solutions on a Number Line

Understanding how to visually represent solutions to equations and inequalities is crucial for gaining a deeper comprehension of their meanings and implications. Let's start with the foundation: the number line.

Understanding the Number Line

The number line is a simple yet powerful tool for visualizing numbers and their relationships. It is an endless line where each point corresponds to a real number. The number line provides a clear way to see positive and negative values, zeros, and even fractions or decimals. Imagine a straight line with a zero in the center. To the right of zero are positive numbers, increasing as you move further to the right. To the left are negative numbers, decreasing as you move further to the left. This visualization helps us understand not only the order of numbers but also their relative size.

Plotting Points for Equations

Now, plot points for equations. When you solve a linear equation, you often find a specific value or set of values for the variable. For example, if you solve ($x + 2 = 5$), you find that ($x = 3$). To plot this solution on the number line, simply locate the point that corresponds to 3 and mark it. By doing so, you clearly indicate that 3 is the solution to the equation. For more complex equations with multiple solutions, such as ($x^2 - 4 = 0$), which simplifies to ($x = 2$) or ($x = -2$), you would plot both 2 and -2 on the number line.

Representing Inequalities Graphically

Representing inequalities graphically follows a similar approach but with additional considerations. Rather than specifying a unique value, inequalities delineate a spectrum of permissible quantities. For example, if you have ($x > 3$), you are looking for all numbers greater than 3. To show this on the number line, you would draw an open circle at 3 (to indicate 3 is not included) and shade the line to the right to show all numbers greater than 3. If the inequality were ($x \geq 3$), you would use a closed circle at 3 to include it and then shade to the right. Similarly, for ($x < -1$), you place an open circle at -1 and shade the line to the left. Graphing inequalities helps visualize the solution set, making it easier to understand and interpret the results.

Real-World Applications of Equations

Knowing how to solve linear equations and inequalities is not just an academic exercise; these concepts have practical applications in everyday life, as we've seen.

Budgeting With Equations

Budgeting is a common activity where linear equations can be particularly helpful. When you set up simple equations, you can gain better control over your finances. For example, if you know your monthly income and fixed expenses, you can use a linear equation to find out how much money you have left for discretionary spending or savings.

Let's say your monthly income is $3000, and your fixed expenses (rent, utilities, groceries) total $1800. You want to save $500 each month. What remains for other expenditures? Here's a basic setup:

- [3000 - 1800 - 500 = x] [3000 - 2300 = x] [x = 700]

So, you have $700 left for other activities like dining out, entertainment, or unexpected expenses. This kind of straightforward calculation helps you understand where your money goes and how to adjust your budget accordingly.

Additionally, equations can be used to plan for future expenses. Suppose you need to save up for a large purchase, such as a new laptop costing $1200. If you save $100 per month, setting up an equation can help you figure out how long it will take to reach your goal:

- [100x = 1200] [x = {1200}÷{100}] [x = 12]

It will take you 12 months to save $1200 at a rate of $100 per month. Understanding these basics can provide clarity and confidence in managing personal finances.

Teaching Support for Children

Another important aspect of understanding linear equations and inequalities is the ability to support your children's education. As a parent, you'll want to help your children with their math homework effectively. Knowing how to solve these problems yourself allows you to guide them through their assignments and demonstrate problem-solving techniques.

When helping children with homework, it's essential to break down the problems into smaller, more manageable steps. For example, if they face an equation like:

[2x + 3 = 11]

Guide them through the process:

1. Subtract 3 from both sides: [2x + 3 - 3 = 11 - 3] [2x = 8]

2. Divide both sides by 2: [{2x}÷{2} = {8}÷{2}] [x = 4]

Explain each step clearly and encourage them to practice similar problems to reinforce their understanding. Additionally, using real-life examples, like budgeting their allowance, can make the learning experience more engaging and relatable.

You can also include creating a structured study environment, using educational resources like online videos or math games, and encouraging a positive attitude towards learning math. Consistent practice and patience are key to helping children become proficient in solving linear equations.

Diagnostic Quizzes on Linear Equations and Inequalities

1. Solve the linear equation: 3x + 7 = 25

2. Solve the linear equation with variables on both sides: 5x - 3 = 2x + 9

3. Solve the linear inequality: 2x - 5 > 11

In this chapter, we've gone through the essential techniques for solving linear equations and understanding inequalities, providing you with fundamental skills that are valuable both academically and in real-world contexts. Once you master how to graph solutions on a number line, you can visualize equations and inequalities clearly, making complex concepts more tangible. Whether it's figuring out budget constraints, planning projects efficiently, or assisting your children with their math homework, these visual tools help break down abstract ideas into understandable parts.

Remember, these mathematical concepts aren't just confined to textbooks; they have practical applications that can simplify and enhance various aspects of your daily life. From managing finances and planning tasks to making informed decisions at work and supporting educational activities at home, the ability to solve linear equations and understand inequalities is immensely beneficial. Keep practicing these techniques and applying them to real-life scenarios to build confidence and proficiency in your math skills.

CHAPTER 12

Introduction to Graphing

Graphing is a fundamental skill in mathematics that opens up a world of understanding by visually representing relationships between variables. When you start plotting points on the Cartesian coordinate system, it's like deciphering a map that reveals how different elements interact and change. This chapter is your guide to mastering the art of graphing linear equations, a type of equation where the graph is a straight line. After you get a firm grasp of this concept, you'll know how to work out more complex mathematical problems.

In this chapter, we will first break down what a linear equation is and how it can be expressed in the form $y = mx + b$. Then, we will look at, identify, and interpret the slope and y-intercept—key components that define the behavior of these equations on a graph. From there, we'll guide you through the step-by-step process of plotting a linear equation, starting from identifying the slope and y-intercept to drawing the line on the graph. We'll also see how you can interpret these graphs to understand what they tell us about the relationship between x and y. Practical examples and exercises will help solidify these concepts, making them second nature.

Plotting Linear Equations

When we talk about graphing linear equations, we're looking at one of the fundamental skills in mathematics. This section will help you become familiar with plotting these equations on a graph so you can better understand their relationships visually.

Definition of Linear Equations

To start, let's clearly define what a linear equation is. A linear equation can be expressed in the form y = mx + b, where "y" is the dependent variable, "x" is the independent variable, "m" represents the slope of the line, and "b" signifies the y-intercept. The slope "m" indicates how steep the line is, while the y-intercept "b" shows where the line crosses the y-axis. This form of equation is relevant because it directly links algebraic expressions to their graphical representations.

Graphic Steps

Now that we've got the basics down let's move on to the steps for graphing a linear equation on the Cartesian plane. Graphing a linear equation involves a few essential steps:

1. **Identify the Slope and Y-Intercept:** In the equation y = mx + b, determine your slope (m) and y-intercept (b). For instance, if your equation is y = 2x + 3, the slope (m) is 2, and the y-intercept (b) is 3.

2. **Plot the Y-Intercept:** Start by plotting the y-intercept on the graph. This point lies on the y-axis. For our example, y = 2x + 3, plot the point (0, 3) on the y-axis.

3. **Use the Slope to Find Another Point:** From the y-intercept, use the slope to find another point on the graph. Remember, the slope is rise over run. With a slope of 2, you rise 2 units up and run 1 unit to the right. From (0, 3), you reach the point (1, 5).

4. **Draw the Line:** Once you have two points, draw a straight line through them, extending in both directions. This line represents the solutions to the equation y = 2x + 3.

Interpreting Graphs

Having a good grasp of how to interpret the graphs of linear equations is equally important. When you look at a graph, several key details can be interpreted:

- **Slope:** The slope tells you how quickly y changes as x changes. A steep slope means y changes rapidly. If the slope is positive, the line moves upward; if negative, it slopes downward.

- **Y-Intercept:** This is where the line crosses the y-axis. It indicates the value of y when x is zero.

- **X-Intercept:** It might also be useful to identify where the line crosses the x-axis, which happens when y is zero.

Interpreting these elements accurately allows you to gain insights into the relationship between x and y. For example, if you're analyzing financial data, understanding the slope could tell you how quickly profits are increasing relative to revenue.

To solidify these concepts, practice is essential. Let's look at some real-life examples and exercises:

Understanding and Calculating Slope

Slope, often represented by the letter "m," measures the steepness and direction of a line on a Cartesian coordinate system. Think of it as a way to describe how tilted or flat a line is. In simple terms, the slope tells us how much the y-coordinate (vertical change) changes for every unit change in the x-coordinate (horizontal change). A positive slope means the line tilts upward from left to right, while a negative slope indicates the line tilts downward.

Calculating Slope

Now, let's have a look at how we can calculate the slope. One common method involves using two points on a line, say (x_1, y_1) and (x_2, y_2). The formula to calculate the slope between these two points is $(y_2 - y_1) \div (x_2 - x_1)$. This ratio gives us the rate at which the y-value changes per unit increase in the x-value. It's helpful to break this down with an example. Suppose you have two points on a line: (2, 3) and (5, 7). Plugging these values into our formula, we get $(7 - 3) \div (5 - 2) = 4 \div 3$. Therefore, the slope of the line passing through these points is 4/3, meaning for every 3 units you move horizontally, the vertical value increases by 4 units.

Calculating the slope from an equation is another method. For linear equations in the form $y = mx + b$, the coefficient of x (represented by m) denotes the slope. If your equation is $y = 2x + 1$, for example, the slope is 2. This method is straightforward but very effective for quickly identifying the steepness of a line without any graphing involved.

Slope in Context

The concept of the slope goes beyond mathematical equations; it has real-life applications that make understanding it all the more vital. Take, for instance, the case of a business owner tracking revenue over time. If the revenue data is plotted on a graph, the slope of the line connecting these data points can tell the business owner how quickly revenue is increasing or decreasing. Similarly, engineers use slope calculations to design roads, ramps, and other structures. They must ensure that the slope is optimal for safety and functionality.

Consider the example of driving a car up a hill. The slope of the hill affects how much power the car needs to ascend. A steeper slope requires more engine power, whereas a gentler slope is easier to climb. Understanding these principles can be incredibly useful in everyday decision-making, whether planning a route or designing infrastructure.

1. Practice is key to solidifying your understanding of slopes. Let's start with an exercise:

2. Find the slope of a line that passes through the points (6, 8) and (10, 14).

3. Using the slope formula, we get (14 - 8) ÷ (10 - 6) = 6 ÷ 4, which simplifies to 3/2. And so the slope is 3/2.

Another practice problem could involve an equation. Given the equation $y = -3x + 5$, identify the slope. Here, the slope is simply -3, as it's the coefficient of x.

Interpreting slopes in various contexts will also help cement your understanding. For example, in the stock market, the slope of a trend line on a stock chart can indicate the rate at which stock prices are rising or falling. Analyzing these slopes can help investors make informed decisions about buying or selling stocks.

One engaging exercise to further reinforce this concept involves creating your own graphs. Plot different sets of points and calculate the slopes. Then, verify your calculations by drawing the lines and visually checking if they match your computed slopes. This not only helps in understanding but also builds confidence in your graphing skills.

Another useful exercise is to take real-world data, such as daily temperatures, and plot them on a graph. Calculate the slope to determine the rate of temperature change over time. This kind of practical application makes abstract concepts tangible and relevant to everyday life.

Finally, consider working with more complex problems involving piecewise functions, where different segments of the graph may have different slopes. Such exercises challenge you to apply your knowledge in varied scenarios, enhancing your overall comprehension and readiness for more advanced studies or professional applications.

Diagnostic Quizzes on Graphing

1. Plot the following points on a coordinate plane and connect them with a line: A(2, 3) and B(5, 9). Calculate the slope of this line.

2. Find the slope of the line passing through the points (-1, 4) and (3, -2).

3. Calculate the slope of the line represented by the equation $y = 2x + 5$.

Reflection Questions on Intermediate Algebra

Working With Algebraic Expressions

1. Find the GCF and factor it out: $24x^3y^2 - 36x^2y^3$

2. Recognize and factor the difference of squares: $49a^2 - 64b^2$

3. Factor the following quadratic expression: $x^2 - 7x + 12$

Solving Linear Equations and Inequalities

1. Solve the linear equation: $4x - 9 = 23$

2. Solve the linear equation with variables on both sides: $3x + 5 = x - 7$

3. Solve the linear inequality: $2x + 5 < 17$

Plotting Linear Equations and Calculating Slope

1. Plot the following points on a coordinate plane and connect them with a line: A(-1, 2) and B(3, 6). Calculate the slope of this line.

2. Find the slope of the line passing through the points (0, -3) and (4, 5).

3. Calculate the slope of the line represented by the equation $y = -2x + 3$.

We've covered everything from identifying the slope and y-intercept to plotting points and drawing lines. These skills form the foundation for interpreting graphs and understanding relationships between variables. Whether you're analyzing financial data, tracking changes over time, or simply solving math problems, being able to graph and interpret linear equations is a valuable tool.

As you move forward, remember that practice is key to mastering these concepts. Try plotting different equations and interpreting their graphs to build your confidence. You'll find that with each exercise, these principles become more intuitive and easier to apply in real-world scenarios. Keep at it, and soon, you'll be tackling more advanced mathematical topics with ease!

CHAPTER 13

Financial Math

Exploring the world of financial math might seem, at first glance, a little complicated, but it's an essential skill for managing your money effectively. It can come in handy if you're taking out a loan, investing in a savings account, or planning your budget. This is because understanding these important mathematical concepts will help you make informed decisions that benefit your financial health.

In this chapter, we'll go into the specifics of simple and compound interest calculations and see how they relate to real-world financial situations. You will get the knowledge to calculate the extra expenses and profits linked to investments and loans. We will also discuss personal budgeting strategies to assist you in controlling your monthly earnings and expenses. You'll be better prepared to manage financial opportunities and obstacles if you understand these ideas, which will put you on the path to greater financial success and stability.

Simple and Compound Interest Calculations

When discussing financial math, a fundamental concept to grasp is simple interest. Let's start with this.

Understanding Simple Interest

Simple interest is the additional amount paid on a principal sum at a set rate over time. The formula for calculating simple interest is I = P x r x t.

Here, 'I' stands for interest, 'P' represents the principal (the initial amount of money), 'r' is the interest rate, and 't' denotes time. This straightforward formula allows adults to make informed decisions about borrowing and lending money.

If someone borrows $1,000 at an annual interest rate of 5% for three years, the total interest owed would be: I = 1000 x 0.05 x 3, which equals $150.

Knowing this is important because it affects everyday financial scenarios. Understanding how to calculate simple interest ensures that you are aware of the total cost of borrowing or the potential gains from saving. It also helps in comparing different financial products to determine the best option based on your needs.

Exploring Compound Interest

However, while simple interest calculations are essential, many financial instruments operate on compound interest, which adds another layer of complexity but also offers greater potential for growth. In compound interest calculations, growth occurs not only on the original capital sum but also on the interest accrued during preceding time intervals. The formula used to calculate compound interest is A = P(1 + r÷n)^(nt), where 'A' stands for the amount of money accumulated after n periods, including interest, 'n' is the number of times interest is compounded per unit time period, and the other variables remain as previously defined.

Practical Applications of Interest

Let's consider a practical example to understand the impact of compound interest. Suppose you invest $1,000 in a savings account that offers an annual interest rate of 5%, compounded quarterly.

Using the compound interest formula, $A = 10,00(1 + 0.05 \div 4)^{(4 \times 3)}$. After plugging in the numbers, we get approximately $1,161.62 after three years. As you can see, compound interest results in more substantial growth compared to simple interest, making it a powerful tool for accumulating wealth over time.

But let's look at a real-life scenario. Imagine you're comparing two savings accounts: one offering simple interest and the other providing compound interest. Knowing how to use these formulas lets you evaluate which account will yield higher returns over your desired investment period. In the same way, when considering mortgage options, understanding compound interest gives you insight into how much you'll end up paying over the life of the loan. These practical applications highlight the importance of mastering both simple and compound interest calculations.

Another relevant area where understanding interest calculations comes into play is investments. Evaluating investment returns often involves comparing simple interest versus compound interest earnings. For example, you might be deciding between a bond that pays simple interest and a mutual fund account that offers compound interest. When you apply the respective formulas, you can predict which option aligns with your financial goals and risk tolerance.

Comprehending these concepts also prepares you to spot common mistakes and avoid pitfalls associated with interest calculations. One frequent error is misunderstanding the difference between simple and compound interest. This mistake can lead to inaccurate assessments of financial products, potentially resulting in less favorable financial decisions. For example, some people mistakenly believe that a higher interest rate always guarantees better returns without considering whether the interest is simple or compounded.

Personal Budgeting Techniques

Effective personal budgeting is a crucial skill for managing your finances better. Crafting an efficient budget requires understanding your income, expenses, and savings goals. Let's explore a step-by-step guide to creating a budget.

Creating a Budget

Creating a budget begins with calculating your total monthly income. This includes your salary, any side income, freelance work, or other regular sources of money. Knowing your exact income helps you understand what you have to work with each month.

Next, compile a list of all your monthly expenses. Start with fixed expenses like rent or mortgage payments, utilities, insurance premiums, and loan repayments. These are non-negotiable and must be paid every month. Then, add variable expenses such as groceries, transportation, entertainment, dining out, and miscellaneous spending. Variable expenses can fluctuate, so it's essential to track them carefully.

Once you have a comprehensive list of your expenses, subtract the total from your income. The amount left is your disposable income. Ideally, some portion of this should be allocated toward savings goals. If your expenses exceed your income, you should identify areas where you can cut back.

A practical tool for simplifying this process is the 50/30/20 rule. This budgeting method suggests dividing your after-tax income into three categories: 50% for needs, 30% for wants, and 20% for savings. Needs include essentials like housing, groceries, and healthcare. Wants are discretionary expenses like dining out, hobbies, and travel. The remaining 20% should go towards savings or paying off debt. Using this rule can help you balance necessary spending with saving and enjoying life.

Templates and budgeting apps can further streamline the budgeting process. Many free and paid options are available that allow you to input your income and expenses, providing a visual aid to track your spending. These tools can remind you of upcoming bills and alert you when you're overspending in certain categories.

Monitoring and Adjusting Budgets

Regularly reviewing and adjusting your budget is vital as life changes. Income can vary due to job changes, bonuses, or pay raises, and expenses may increase with new financial responsibilities like a child or a car purchase. It's important to revisit your budget monthly to ensure it still fits your current financial situation.

Problems commonly faced while sticking to a budget include unexpected expenses like medical bills or car repairs. To manage these surprises, it's wise to build an emergency fund. This fund is separate from your savings account with enough money to cover at least three to six months' worth of living expenses. This cushion can provide peace of mind and financial security during unforeseen events.

Strategies for Reducing Expenses

Identifying areas to cut costs is another thing you should take into consideration if you're looking for an effective budget. Start by categorizing your spending into needs and wants. For example, do you need that daily coffee shop visit, or could you make coffee at home? Small changes like cooking at home more often, canceling unused subscriptions, or shopping during sales can accumulate significant savings over time.

Practical examples can illustrate effective cost-saving measures. For instance, switching to a cheaper phone plan, buying generic brands instead of name brands, and using public transportation instead of driving can save money. Another tip is to negotiate bills and services; sometimes, companies offer discounts or better rates if you ask.

Setting Saving Goals

Automating your savings is a powerful strategy. Setting up automatic transfers from your checking account to your savings account ensures you consistently set aside money each month without having to think about it. This method makes saving a habit rather than an afterthought.

When it comes to creating achievable savings goals, start with identifying what you're saving for. Whether it's a vacation, a down payment on a house, or building an emergency fund, clear goals can motivate you to stick to your budget. Next, determine how much you need to save and by when. Divide this amount by the number of months until your goal date to find out how much you need to save each month. Adjust your budget to include this monthly saving amount, and track your progress regularly.

Having an emergency fund is not just important but essential. Life is unpredictable, and financial setbacks can happen at any time. If you set aside money specifically for emergencies, you safeguard yourself against the stress and hardship of unforeseen expenses. Prioritize building this fund even if it means cutting back on non-essentials temporarily.

Diagnostic Quizzes on Financial Math

1. You invest $5,000 at a simple interest rate of 4% per year for 3 years. How much interest will you earn?

2. If you invest $2,000 at 5% interest compounded annually for 4 years, what will be the final amount?

3. Calculate the final amount if $3,000 is invested at 6% per year, compounded quarterly for 2 years.

This chapter covered the fundamental ideas of compound and simple interest as well as their applications in a range of financial situations. You'll be more capable of making judgments about investing, borrowing money, and saving money if you know how to compute both types of interest. While compound interest illustrates the possibility for exponential development in your assets, simple interest gives you a clear picture of what you owe or potentially earn over time.

We also looked into personal budgeting techniques, emphasizing the importance of tracking income and expenses to achieve financial stability. Creating a budget, using tools like the 50/30/20 rule, and setting up an emergency fund are all practical steps you can take to manage your money wisely. When you master these skills, you'll not only improve your own financial health but also be prepared to help others, whether it's guiding your kids with their homework or assisting colleagues in a professional setting. In the next chapter, we will be exploring statistics and probabilities.

CHAPTER 14

Statistics and Probability

Statistics and probability serve to make sense of data in our everyday lives and professional landscapes. Once you get a good grasp of how to calculate and interpret statistical measures like mean, median, and mode, you'll know how to better summarize vast datasets and spot patterns that could otherwise go unnoticed.

Here, we will introduce you to the basics of calculating mean, median, and mode and explain their significance in different contexts. Also, how the mean provides an average but can be skewed by outliers, why the median offers a more accurate middle point in asymmetrical datasets, and how the mode helps identify recurring trends. You'll also get a primer on basic probability concepts, understanding not just theoretical probabilities but also those based on experiments and personal judgments.

Calculating Mean, Median, and Mode

These measures of central tendency highlight different aspects of a dataset and help make sense of large amounts of information quickly.

Mean: The Power of Averaging

The mean, often referred to as the average, is calculated by summing all the values in a dataset and then dividing by the number of values. It provides a straightforward way to summarize data, making it invaluable in reports and presentations. For example, if you're tracking your weekly expenses, adding up all costs and dividing by seven gives you your average daily expense. This method helps pinpoint typical spending patterns and budgets more effectively.

However, it's important to note that the mean can be influenced by extreme values, known as outliers. If one of your daily expenses was unusually high, say due to an emergency, the mean might not represent your ordinary spending accurately. In such cases, relying solely on the mean could lead to skewed interpretations. Thus, while the mean offers a quick snapshot, it's crucial to consider its sensitivity to outliers in decision-making processes.

In practice, businesses extensively use the mean to analyze various metrics, from employee performance to customer feedback scores. For instance, calculating the average rating of a product can help companies assess overall customer satisfaction and make necessary improvements.

Median: Finding the Middle Ground

The median represents the middle value in a sorted dataset. To find the median, you sort the numbers from the lowest to the highest and pick the middle one. When the dataset contains an even quantity of elements, the median value is computed by taking the arithmetic mean of the two middle figures. This measure is particularly helpful in datasets with skewed distributions or outliers since it isn't affected by extremely high or low values, unlike the mean.

For example, consider income distribution in a neighborhood. If most residents earn between $40,000 and $60,000 annually, but a few earn millions, the mean income would be disproportionately high, suggesting greater wealth across the board than what most people actually experience. The median income, on the other hand, would provide a truer picture of the typical earnings, giving better insights for policy-making or social analysis.

Real-life scenarios often demand the use of medians over means for accuracy. In education, median test scores are used to understand the performance of students without the distortion caused by exceptionally high or low scores. Similarly, in real estate, median home prices offer prospective buyers a realistic idea of property costs.

Mode: Spotting Common Trends

The mode is the value that appears most frequently in a dataset. Unlike mean and median, the mode doesn't necessarily point to the center but rather highlights the most common occurrence. Identifying trends becomes easier with mode, especially when frequency matters more than the numerical average.

Take market research as an example. If a fashion retailer wants to know which clothing size is most popular, the mode will reveal this insight directly. If a clothing store stocks more of the modal size, they can meet customer demand more efficiently. Additionally, in healthcare, understanding the most common symptoms reported by patients helps in creating effective treatment plans and allocating resources appropriately.

While useful, the mode has limitations. Some datasets may have multiple modes or none at all, making them less reliable in certain analyses. Hence, it should be used judiciously alongside other measures of central tendency.

Comprehensive Insights: Using Measures Together

To gain a well-rounded understanding of any dataset, it's beneficial to use mean, median, and mode together. Each measure offers unique insights, and their combined usage can illuminate different facets of the data.

Consider a company analyzing employee salaries. The mean salary provides an overall average, which can be useful for budget planning. The median salary shows the typical earning level, highlighting what most employees might expect. Meanwhile, the mode can indicate the most common salary bracket, helping the company understand salary distribution patterns.

When comparing these measures, it's easier to detect anomalies and get a clearer picture of the dataset's characteristics. For instance, a significant difference between mean and median salaries might suggest income inequality within the organization, prompting further investigation.

Understanding when to use each measure is just as important. While the mean is ideal for symmetric distributions without outliers, the median suits skewed distributions, and the mode excels in categorical data analysis. This discernment ensures that decisions based on data are well-informed and accurate.

Basic Probability Concepts

Probability is a fascinating but often misunderstood concept, crucial to both day-to-day decision-making and many professional fields. Probability is the measure of the likelihood that an event will occur, usually expressed as a fraction or decimal between 0 and 1. For instance, a probability of 0 indicates an impossible event, whereas a probability of 1 signifies a certainty. To illustrate, if you flip a coin, the probability of landing heads is 0.5, reflecting that out of two possible outcomes, one outcome is favorable.

Definition of Probability

Understanding probability can help significantly in various aspects of life. One of these real-life examples is taking risk assessments. Professionals in sectors such as finance and insurance often rely on probability to calculate potential losses and make informed decisions. Even in daily life, we use probability without realizing it—

whether it's deciding to carry an umbrella based on a weather forecast or estimating the chances of winning a game.

Types of Probability

There are different types of probability: theoretical, experimental, and subjective. Theoretical probability is based on reasoning or intuition. For example, the theoretical probability of rolling a three on a six-sided die is 1/6 because there is one favorable outcome out of six possible outcomes.

Experimental probability, on the other hand, relies on actual experiments or historical data. If you flip a coin 100 times and it lands on heads 48 times, the experimental probability of getting heads would be 48/100, or 0.48. This type offers more practical insights as it's based on observation and experimentation rather than pure logic.

Subjective probability deals with personal judgment or experience. For instance, a sports analyst predicting a team's chances of winning based on their performance is using subjective probability. While it might incorporate some statistical analysis, it largely depends on the individual's expertise and intuition.

These types of probability are not merely academic; you can apply them to real life. Marketing professionals use them to predict consumer behavior, financial analysts assess investment risks, and healthcare providers evaluate treatment effectiveness. Realizing these differences can significantly boost your ability to apply these concepts appropriately across various scenarios.

Rules of Probability

Moving on, understanding the rules of probability is vital for combining probabilities of independent events. One fundamental rule is the addition rule, which helps in calculating the probability of either of two mutually exclusive events occurring. For example, if you want to know the probability of drawing an ace or a king from a deck of cards, you add their individual probabilities: 4/52 + 4/52 = 8/52, or approximately 0.154.

The multiplication rule is another basic principle used to find the probability of two independent events occurring together. If you roll two dice and you're curious about the probability of both showing sixes, you multiply the probability of each event: $1/6 \times 1/6 = 1/36$, which is roughly 0.028.

These basics are essential when dealing with compound events, where multiple probabilities interact. They find practical applications in numerous fields, including business forecasting and even planning social events. Practicing these calculations reinforces key concepts, helping to avoid common pitfalls like misinterpreting independent and mutually exclusive events. Misunderstanding these can lead to errors, such as wrongly assuming that the probability of rolling two sixes in a row is high just because six came up recently.

Diagnostic Quizzes on Statistics and Probabilities

1. Calculate the mean, median, and mode for the following dataset: 7, 9, 4, 6, 7, 10, 3, 7, 8, 5.

2. In a bag, there are 5 red marbles, 3 blue marbles, and 2 green marbles. If you draw one marble at random, what is the probability of drawing either a red or a green marble?

3. A student must answer 3 out of 5 questions on a quiz. What is the probability of the student selecting questions 1, 2, and 3 in that order?

Reflection Questions on Real-Life Math Applications

Financial Math

1. Calculate the simple interest on a principal of $2,500 at 4.5% per annum for 3 years.

2. If $3,000 is invested at 6% per annum compounded annually for 4 years, what will be the final amount?

3. Compare the final amounts after 2 years for an investment of $5,000 at 5% per annum: a) With simple interest; b) Compounded quarterly.

Statistics and Probabilities

1. Calculate the mean, median, and mode for the dataset: 14, 18, 22, 14, 16, 18, 20, 22, 14, 18.

2. In a standard deck of 52 cards, what is the probability of drawing: a) A heart or a king? b) A red card or an ace?

3. If the probability of rain tomorrow is 0.4 and the probability of wind is 0.3, what's the probability of both rain and wind, assuming the events are independent?

The fundamental concepts of statistical measures have been reviewed in this chapter, with an emphasis on the calculation and assessment of the mean, median, and mode. These tools are essential for interpreting statistics. Making better judgments may be simplified by being aware of the unique insights that each measure offers into a dataset. This will enable you to see trends and patterns. Remember that the mean gives you an average, the median shows you the middle value, and the mode tells you about the most frequent occurrence in your data.

Additionally, we introduced some basic probability concepts that can greatly enhance your decision-making skills. Knowing the likelihood of various outcomes helps in assessing risks and predicting future events. From flipping a coin to evaluating stock market trends, probability plays a crucial role in both everyday life and professional settings. And this is all you need to know to have a good grasp of fundamental mathematics. In the next chapter, we will give you a quick but comprehensive review of what we learned in this book so you can solidify your knowledge.

CHAPTER 15

Comprehensive Review

This chapter will help you understand how you can identify your weaknesses and other strategies to improve your mathematical skills. This isn't just about remembering formulas but about seeing how different parts of math fit together. Think of it as assembling a puzzle where each piece, once placed properly, gives a clearer picture of your overall understanding.

Problem Sets Focused on Difficult Areas

Understanding mathematical concepts is not always straightforward. Many adults and professionals looking to refresh their skills often find there are gaps in their understanding that need addressing. And here, we will focus on identifying and reinforcing those essential math concepts where you might struggle and making sure you have a well-rounded mastery of the subject.

Identify Areas of Weaknesses

First and foremost, identifying weaknesses is crucial. You will start by completing diagnostic exercises specifically designed to highlight areas needing more attention, such as the ones in the different chapters we've covered. These activities aren't meant to be daunting but rather serve as a tool for self-discovery. Once you recognize which topics cause difficulty, whether it's algebra, geometry, or calculus, you can focus your efforts more effectively. For example, you might find fractions persistently tricky. Diagnostic testing will pinpoint this issue, allowing you to dedicate more time to practice, thereby boosting your overall confidence and skill level. Acknowledging weaker areas isn't about highlighting failure; it's about paving the way for improvement and developing a focused strategy for learning.

Comprehensive Review Exercises

Once weaknesses are identified, engaging in comprehensive review exercises is the next step. These are varied problems crafted to challenge and reinforce what has been learned. A mix of simple and complex questions across different topics ensures that you are not only reviewing but also retaining the information. The advantage here is twofold: you broaden your exposure to different problem types, which solidifies your understanding, and you move beyond rote memorization to being able to apply concepts fluidly in different contexts. To give you an example, solving multiple types of word problems not only enhances your problem-solving skills but also prepares you for real-world applications, such as calculating interest rates or budgeting for projects. Comprehensive review exercises reduce anxiety by making unfamiliar questions feel familiar through consistent practice.

Peer Review Discussion

Another powerful method for mastering math concepts is peer review discussion. Sharing solutions with peers offers an opportunity for collaborative learning. When you explain your reasoning to others, it deepens your own understanding while exposing you to different problem-solving strategies. Engaging in these discussions can turn a solitary study session into an interactive one and cultivate a supportive learning environment. So, if you're struggling with a particular concept, a peer may offer a different perspective or a simpler method that could make things click.

Self-Assessment Reflections

Self-assessment reflections provide the final layer of reinforcement. After completing problem sets, taking the time to reflect on performance is essential. Self-assessment allows you to recognize your improvements and understand your mistakes better. It's about creating a positive feedback loop where each success builds upon the last, and each mistake is seen as a learning opportunity. Documenting progress helps in visually seeing growth over time, which can be incredibly motivating. For instance, keeping track of how your scores improve on diagnostic tests or noting down common errors and working on them can lead to significant long-term benefits. Self-reflection also instills a sense of personal accountability where you take charge of your educational journey.

Timed Drills for Speed and Accuracy

Structured timed drills are an effective method for developing speed and accuracy in mathematical problem-solving, essential for both test-taking and practical real-world applications. Implementing this strategy involves setting time constraints, incrementally increasing difficulty, tracking progress, and encouraging the development of personalized strategies.

Setting Time Constraints

Introducing a series of timed drills simulating real testing scenarios can significantly enhance decision-making speed. These exercises push individuals to think quickly under pressure, simulating actual exam conditions, which is crucial for standardized tests and professional assessments. When you set strict time limits on various math problems, you can improve not only your speed but also your intuitive understanding of mathematical relationships.

Incremental Difficulty

Starting with simpler problems before progressing to more complex ones helps cater to all skill levels. Incremental difficulty ensures that you build confidence as you go along. Initially, the focus could be on single-step problems involving basic operations like addition and subtraction. Once comfortable, the complexity can gradually increase to multi-step problems or those requiring higher-order thinking, such as algebraic equations or geometry problems. This progressive challenge allows you to master fundamental concepts slowly, preventing feelings of overwhelm and fostering a sense of achievement regularly. Over time, the gradual escalation in difficulty can help bridge gaps in knowledge and build a strong foundation necessary for tackling advanced topics confidently.

Tracking Progress

Equally important is maintaining records of times and accuracy rates for each drill to monitor improvement. Keeping track of performance provides tangible evidence of progress, helping to identify trends and areas needing further practice. You might notice your speed in solving multiplication problems improves over several sessions, while your accuracy in fractions needs more work. This self-monitoring can improve personal growth and accountability, making the learning process more transparent and manageable. Also, seeing consistent improvements, no matter how small, can serve as a powerful motivator, encouraging continued effort and dedication.

Strategic Development

Encouraging the development of personalized strategies is another vital aspect of enhancing speed and accuracy. You should be guided to develop your own methods for quickly identifying solutions. This can be practicing mental math techniques, recognizing common patterns, or using shortcuts. For example, when faced with complex calculations, breaking problems down into smaller, more manageable steps can save time and reduce errors. Experimenting with different approaches promotes ownership of learning as you discover what works best for you, leading to more efficient and effective problem-solving techniques.

You can also benefit from reflecting on your strategies and adapting them as needed. After completing a set of drills, take a few minutes to review which methods were successful and which ones need adjustment. This can provide valuable insights. It's a reflective practice that can lead to continuous improvement as you refine your approaches based on real-time feedback.

Suggested Tools

There are many tools out there that you can use to solidify your knowledge and even improve it. Let me give you a few examples.

GeoGebra (geogebra.org)

- Free, open-source software for mathematics

- Features: Interactive geometry, algebra, statistics, and calculus

- Great for visualizing graphs, geometric constructions, and function transformations

- Available as web-based and desktop applications

Mathway (mathway.com)

- Problem-solving tool covering various math topics

- Features: Step-by-step solutions, graphing calculator

- Useful for checking work and understanding solution processes

- Offers a mobile app for on-the-go problem-solving

Wolfram Alpha (wolframalpha.com)

- Computational knowledge engine

- Features: Advanced calculations, data analysis, and visualization

- Excellent for complex mathematical operations and data representation

- Provides detailed explanations and additional related computations

Khan Academy (khanacademy.org)

- Free online learning platform

- Features: Video lessons, practice exercises, and personalized learning dashboard

- Covers math from elementary to college level

- Includes other subjects like science, economics, and computer programming

CHAPTER 16

Final Practice Test

Taking a final practice test is an excellent way to consolidate your math skills and gauge your understanding. This chapter offers a comprehensive test that covers all the concepts presented throughout the book, providing you with a unique opportunity to apply what you've learned in various scenarios. The test is designed not only to challenge your capabilities but also to make the learning process engaging through a variety of problem types and difficulties.

In this chapter, you'll encounter a diverse set of problems targeting key areas such as arithmetic, algebra, geometry, and financial math. Each section is crafted to reflect real-world applications, ensuring you see the practical value of your mathematical knowledge. With questions ranging from easy to difficult, you'll find suitable challenges for your current proficiency level, helping you identify strengths and areas needing improvement. Additionally, the chapter emphasizes important problem-solving strategies, guiding you on how to approach different kinds of problems systematically. This section aims to build your confidence and solidify your mastery of essential math skills.

Practice Test

Numbers and Operations

1. $18 + 3 \times (7 - 2) \div 5 + 2\text{\textasciicircum}2$

2. $24/4 \times 3 - 5 + 2\text{\textasciicircum}3$

3. $5 + 3 \times (4 - 2)\text{\textasciicircum}2 - 6/2$

Fractions and Decimals

1. Add the following and express your answer as a decimal: $3/8 + 0.625$

2. Multiply the following and express your answer as a fraction in its simplest form: $2/3 \times 0.75$

3. Convert 5/6 to a decimal, rounding to three decimal places if necessary.

Percentages and Ratios

1. In a class of 200 students, 45% are boys. How many girls are in the class?

2. A shirt originally priced at $80 is on sale for 15% off. What is the sale price?

3. The ratio of cats to dogs in a pet store is 3:5. If there are 24 dogs, how many cats are there?

Factors and Multiples

1. Find the prime factorization of 84.

2. Calculate the LCM of 18 and 24 using prime factorization.

3. Find all the factors of 72 using its prime factorization.

Basic Algebraic Concepts

1. Simplify the following expression: $3x + 2y - 5x + 4y - 7$

2. Combine like terms in the expression: $2a\text{\textasciicircum}2 - 3ab + 5a\text{\textasciicircum}2 + 2ab - 4a\text{\textasciicircum}2$

3. Simplify the following expression: 4(x + 2) - 2(x - 3) + 3x

Integers and Exponents

1. Evaluate the following expression: (-3) + 7 - (-5) + (-2)

2. Multiply the following integers: (-4) x 3 x (-2)

3. Simplify the following expression using the power of powers rule: (x^3)^4

Proportions and Rates

1. A recipe calls for 2 cups of sugar for every 5 cups of flour. If you want to use 8 cups of flour, how many cups of sugar do you need?

2. A car travels 240 miles in 4 hours. At this rate, how far will it travel in 6 hours?

3. In a school, the ratio of teachers to students is 1:25. If there are 750 students, how many teachers are there?

Geometry Fundamentals

1. A rectangular garden is 15 feet long and 10 feet wide. Calculate its perimeter and area.

2. A circular pond has a radius of 4 meters. Find its circumference and area. (Use π = 3.14)

Shapes and Their Properties

1. A rectangular prism has a length of 8 cm, a width of 5 cm, and a height of 3 cm. Calculate its volume.

2. A cylindrical water tank has a radius of 2 meters and a height of 5 meters. What is its volume? (Use π = 3.14)

3. A sphere has a radius of 6 inches. Calculate its volume. (Use π = 3.14)

Algebraic Expressions

1. Factor the following algebraic expression completely:1 2x^2y - 18xy^2 + 24xy

2. Factor the following quadratic expression: x^2 + 7x + 12

3. Simplify the following polynomial by combining like terms: 3x^3 - 2x^2 + 5x - 4x^3 + 6x^2 - 2x + 8

Linear Equations and Inequalities

1. Solve the linear equation with decimals: 0.5x + 1.2 = 0.8x - 0.6

2. Solve the following absolute value equation: [2x - 5] = 11

3. Solve this linear inequality with a variable denominator: (x + 2)÷(x - 3) ≤ 2

Graphing

1. Calculate the slope of a line that passes through the points (-3, 4) and (2, -1).

2. Find the slope of a line with the equation 3x - 2y = 12.

3. A roof rises 5 feet over a horizontal distance of 20 feet. What is its slope?

Financial Math

1. Problem: A loan of $5,500 is taken out at 4.5% simple interest for 3 years. How much interest will be paid?

2. Problem: $3,000 is invested at 6% per annum, compounded quarterly. What will be the balance after 2 years?

3. Problem: How long will it take for $2,000 to double at 8% simple interest?

Statistics and Probabilities

1. The following data represents the number of books read by 9 students in a month: 3, 5, 2, 4, 3, 6, 2, 4, 5. Calculate the mean, median, and mode.

2. A bag contains 4 red marbles, 3 blue marbles, and 5 green marbles. If two marbles are drawn without replacement, what is the probability that both are green?

3. The weights (in kg) of 7 boxes are 12, 15, 11, 13, 15, 14, and 12. Calculate the mean and standard deviation.

Conclusion

As this mathematics refresher draws to an end, let's pause to consider the road we've traveled together. We started by going over the fundamentals again, such as learning the arithmetic sequence of operations, and then worked our way up to more difficult subjects like financial math. Every chapter was created with the intention of building on the one before it, creating a coherent knowledge of fundamental mathematical concepts. Throughout the process, we studied algebra, geometry, and statistics and even took a go at calculus. Our goal has been to provide you with an extensive arsenal of mathematical ideas that you can employ with certainty in different areas of your life.

Think back to where you began and consider how far you've come. You now possess the ability to solve equations, understand geometric properties, analyze data, and apply mathematical reasoning to everyday situations. These skills are not confined to academic exercises; they have real-world applications, as we've seen multiple times, that can increase both your personal and professional life.

Imagine the confidence you'll feel when faced with real-life scenarios that require mathematical proficiency. Just imagine yourself effortlessly assisting your child with their math homework, guiding them through problems with ease. Or picture successfully managing your household budget and making informed financial decisions based on sound mathematical principles. Consider the advantage you'll have in your profession, where an understanding of critical math concepts can improve your performance and open up new opportunities. The math skills you've honed throughout this book are tools of empowerment, enabling you to go through life's challenges with greater assurance.

Now, it's important to recognize that the journey doesn't end here. While you've acquired a solid foundation, continued practice is key to retaining and strengthening these skills. Make it a habit to revisit the review sections in this book regularly. Set aside a few minutes each week to work on math problems, ensuring that your knowledge remains fresh and applicable. Math is not just a subject to be studied; it's a skill to be exercised.

Don't hesitate to seek out additional resources to deepen your understanding further. There are countless online platforms, courses, and tutorials available that can provide supplementary material and practice exercises. Engaging with these resources will reinforce what you've learned and help you explore new mathematical territories.

As we conclude this book, I encourage you to adopt a growth mindset. Approach math with curiosity and a willingness to overcome obstacles. Don't be discouraged by setbacks; instead, view them as stepping stones toward mastery.

Math is not just about numbers and equations; it's a powerful tool that promotes logical thinking, problem-solving, and analytical skills. It creates a structured approach to tackling complex issues and enhances your ability to think critically. These qualities extend beyond mathematics and can positively impact all areas of your life.

Consider how this newfound confidence in math can influence your future endeavors. If you're considering returning to education, whether it's pursuing a degree or enrolling in certification courses, your strengthened math skills will serve as a solid foundation for success. You'll approach academic challenges with greater self-assurance, knowing that you have the tools to excel. This renewed confidence will propel you forward, opening doors to new opportunities and possibilities.

For those already established in their careers, the practical applications of math go beyond the classroom. In fields such as finance, engineering, and management, mathematical competence is not just a requirement but an asset. Your ability to analyze data, interpret financial statements, and solve complex problems will set you apart and make you a valuable asset to your organization.

Parents, your commitment to refreshing your math skills demonstrates a dedication to supporting your children's education. You've rebuilt your own understanding; you're better prepared to guide them through their studies and create a positive attitude toward math. Your involvement will inspire their confidence, showing them that learning is a lifelong endeavor and that seeking help and improving oneself is a strength, not a weakness. Together, you and your children can embark on a shared journey of discovery and growth.

And, of course, don't forget to practice with the workbook that comes with this book, where you will find many other exercises to continue to hone your mathematical skills.

Solutions

Numbers and Operations

1. 10 + 2 x 3 Step 1: Multiplication before addition 10 + (2 x 3) = 10 + 6 = 16 Answer: 16

2. (8 + 4)÷2 - 3 Step 1: Parentheses first (8 + 4) = 12 Step 2: Division before subtraction 12÷2 - 3 = 6 - 3 = 3. Answer: 3

3. 5^2 + 3 x 4 - 7 Step 1: Exponents first 5^2 = 25 Step 2: Multiplication before addition/subtraction 25 + (3 x 4) - 7 = 25 + 12 - 7 Step 3: Addition and subtraction from left to right 37 - 7 = 30 Answer: 30

Fractions and Decimals

1. Add 5/6 + 7/8: Step 1: Find a common denominator (LCD = 24) 5/6 = 20/24, 7/8 = 21/24 Step 2: Add the numerators 20/24 + 21/24 = 41/24. Step 3: Simplify (if possible) 41/24 = 1 17/24

2. Subtract 3/5 from 7/10: Step 1: Find a common denominator (LCD = 10) 3/5 = 6/10 Step 2: Subtract 7/10 - 6/10 = 1/10

3. Multiply 2/3 by 3/7: Step 1: Multiply numerators and denominators (2 × 3) ÷ (3 × 7) = 6/21 Step 2: Simplify 6/21 = 2/7

4. Divide 9/11 by 3/4: Step 1: Multiply by the reciprocal (9/11) × (4/3) = 36/33 Step 2: Simplify 36/33 = 12/11

5. Add 4 2/3 and 1 5/6: Step 1: Convert to improper fractions 4 2/3 = 14/3, 1 5/6 = 11/6 Step 2: Find a common denominator (LCD = 6) 14/3 = 28/6 Step 3: Add 28/6 + 11/6 = 39/6 Step 4: Simplify and convert to mixed number 39/6 = 6 3/6 = 6 1/2

Percentages and Ratios

1. Discount amount = Original price × Discount percentage = $80 x 0.15 = $12

   Sale price = Original price - Discount amount = $80 - $12 = $68

2. Students playing sports = Total students x Percentage playing sports = 500 x 0.60 = 300 students

   Students playing basketball = Students playing sports x Percentage playing basketball = 300 x 0.40 = 120 students

3. 7/5 = 210/x, where x is the number of girls

   7x = 5 x 210

   7x = 1050

   x = 1050/7 = 150

Factors and Multiples

1. LCM of 18 and 24:

   - 18 = 2 x 3^2

   - 24 = 2^3 x 3

   - LCM = 2^3 x 3^2 = 8 x 9 = 72

2. LCM of 15, 25, and 30:

   - 15 = 3 x 5

- 25 = 5^2

- 30 = 2 x 3 x 5

- LCM = 2 x 3 x 5^2 = 2 x 3 x 25 = 150

3. LCM of 56 and 72:

- 56 = 2^3 x 7

- 72 = 2^3 X 3^2

- LCM = 2^3 x 3^2 x 7 = 8 x 9 x 7 = 504

Reflection Questions on Arithmetic Foundations

Numbers and Operations

1. 18 + 4 x (15 - 3^2)÷3 = 18 + 4 x (15 - 9)÷3 = 18 + 4 x 6/3 = 18 + 24/3 = 18 + 8 = 26

2. 72/9 + 5 x 4 - 12 = 8 + 5 x 4 - 12 = 8 + 20 - 12 = 28 - 12 = 16

3. (36 - 12 x 2) + (8 + 16/4) = (36 - 24) + (8 + 4) = 12 + 12 = 24

Fractions and Decimals

1. 3/8 + 5/12 Common denominator: 24 (3 x 3)÷(8 x 3) + (5 x 2)÷(12 x 2) = 9/24 + 10/24 = 19/24

2. 2/3 x 3/4 = 6/12 = 1/2

3. 0.625 = 625/1000 = 5/8 (simplified)

Percentage and Ratios

1. Boys: 15/40 = 37.5% Girls: 100% - 37.5% = 62.5%

2. 20% of $50 = 0.2 x $50 = $10 Sale price = $50 - $10 = $40

3. 5:3 ratio 5 parts = 24 cats 1 part = 24/5 = 4.8 Dogs = 3 × 4.8 = 14.4 = 14 dogs

Factors and Multiples

1. LCM of 18 and 24: $18 = 2 \times 3^2$; $24 = 2^3 \times 3$; LCM $= 2^3 \times 3^2 = 72$

2. LCM of 36, 48, and 60: $36 = 2^2 \times 3^2$; $48 = 2^4 \times 3$; $60 = 2^2 \times 3 \times 5$; LCM $= 2^4 \times 3^2 \times 5 = 720$

3. LCM of 42 and 70: $42 = 2 \times 3 \times 7$; $70 = 2 \times 5 \times 7$; LCM $= 2 \times 3 \times 5 \times 7 = 210$

Basic Algebraic Problems

1. $3x + 2y - 5x + 4y - 7 = (3x - 5x) + (2y + 4y) - 7 = -2x + 6y - 7$

2. $2(x + 3) - 3(x - 1) = 2x + 6 - 3x + 3 = -x + 9$

3. Like terms in $5x^2 + 3xy - 2x^2 + 4y - xy + 2$:

   A. $5x^2$ and $- 2x^2$

   B. $3xy$ and $- xy$

   C. $4y$ (no like term)

   D. 2 (no like term)

Integers and Exponents

1. $(-3)^4 = 81$; Explanation: $(-3) \times (-3) \times (-3) \times (-3) = 9 \times 9 = 81$

2. $2^5 \times 2^3 = 2^8 = 256$; Explanation: When multiplying powers with the same base, add the exponents.

3. $(x^2)^3 = x^6$; Explanation: When raising a power to a power, multiply the exponents.

Proportions and Rates

1. Let x be the cups of sugar needed 2/5 = x/8; cross-multiply: 2 x 8 = 5x 16 = 5x x = 16/5 = 3.2 You need 3.2 cups of sugar.

2. Let x be the number of boys 3/4 = x/28; cross-multiply: 3 x 28 = 4x 84 = 4x x = 84/4 = 21 There are 21 boys in the class.

3. Let x be the distance traveled in 5 hours 210/3.5 = x/5 Cross-multiply: 210 x 5 = 3.5x 1050 = 3.5x x = 1050/3.5 = 300 The car will travel 300 miles in 5 hours.

Reflection Questions on Pre-Algebra Essentials

Basic Algebraic Concepts

1. 3x + 2y - 5x + 4y - 7 = (3x - 5x) + (2y + 4y) - 7 = -2x + 6y - 7

2. $2a^2$ - 3ab + $5a^2$ + 2ab - $4a^2$ = ($2a^2$ + $5a^2$ - $4a^2$) + (-3ab + 2ab) = $3a^2$ - ab

3. 4(x + 2) - 2(x - 3) = 4x + 8 - 2x + 6 = 2x + 14

Integers and Exponents

1. (-7) + 12 - (-5) + (-3) = (-7) + 12 + 5 + (-3) = 5 + 5 = 10

2. (-4) x 3 x (-2) = (-12) x (-2) = 24

3. 15 - 3 x (-2) + 4÷(-2) = 15 - (-6) + (-2) = 15 + 6 - 2 = 21 - 2 = 19

Proportions and Rates

1. Let x be the cups of flour needed 3/2 = x/5; Cross-multiply: 3 x 5 = 2x 15 = 2x x = 15/2 = 7.5 You need 7.5 cups of flour.

2. Let x be the distance traveled in 6 hours $240/4 = x/6$; Cross-multiply: $240 \times 6 = 4x$ $1440 = 4x$ $x = 1440/4 = 360$ The car will travel 360 miles in 6 hours.

3. First, find the unit rate: $\$3.75 \div 15$ pencils $= \$0.25$ per pencil. Then, multiply by 24: $\$0.25 \times 24 = \$6.00$ 24 pencils will cost $6.00.

Geometry Fundamentals

1. Rectangle: Perimeter $= 2(\text{length} + \text{width}) = 2(12 + 8) = 2(20) = 40$ cm; Area $= \text{length} \times \text{width} = 12 \times 8 = 96$ cm^2

2. Square: Perimeter $= 4 \times \text{side} = 4 \times 15 = 60$ meters; Area $= \text{side}^2 = 15^2 = 225$ m^2

3. Circle: Radius $= \text{diameter}/2 = 14/2 = 7$ inches; Circumference $= 2\pi r = 2 \times 3.14 \times 7 = 43.96$ inches; Area $= \pi r^2 = 3.14 \times 7^2 = 3.14 \times 49 = 153.86$ square inches

Shapes and Their Properties

1. Volume $= 5^3 = 5 \times 5 \times 5 = 125$ cm^3

2. Volume $= 8 \times 6 \times 4 = 192$ cm^3

3. Volume $= 3.14 \times 3^2 \times 10 = 3.14 \times 9 \times 10 = 282.6$ m^3

Reflection Questions on Geometry Basics

1. Perimeter $= 2(\text{length} + \text{width}) = 2(15 + 8) = 2(23) = 46$ cm; Area $= \text{length} \times \text{width} = 15 \times 8 = 120$ cm^2

2. Perimeter $= 10 + 8 + 8 = 26$ m; Area $= (1/2) \times \text{base} \times \text{height} = (1/2) \times 10 \times 6 = 30$ m^2

3. Volume $= 4^3 = 4 \times 4 \times 4 = 64$ in^3

Algebraic Expressions

1. GCF: $6x^2y^2$; Factored: $6x^2y^2(3xy + 4y)$

2. Difference of squares: $(4a + 9b)(4a - 9b)$

3. Factored quadratic: $(x + 7)(x + 4)$

Linear Equations and Inequalities

1. $3x + 7 = 25$ $3x = 18$ $x = 6$

2. $5x - 3 = 2x + 9$ $3x - 3 = 9$ $3x = 12$ $x = 4$

3. $2x - 5 > 11$ $2x > 16$ $x > 8$

Introduction to Graphing

1. Plot A(2, 3) and B(5, 9) on a coordinate plane and connect them. Slope = $(y_2 - y_1) \div (x_2 - x_1) = (9 - 3) \div (5 - 2) = 6 \div 3 = 2$

2. Slope = $(y_2 - y_1) \div (x_2 - x_1) = (-2 - 4) \div (3 - (-1)) = -6 \div 4 = -3/2$

3. In the equation $y = 2x + 5$, the coefficient of x is the slope. Therefore, the slope is 2.

Reflection Questions on Intermediate Algebra

Working With Algebraic Expressions

1. GCF: $12x^2y^2$; Factored: $12x^2y^2(2xy - 3y)$

2. Difference of squares: $(7a + 8b)(7a - 8b)$

3. Factored quadratic: $(x - 3)(x - 4)$

Linear Equations and Inequalities

1. $4x - 9 = 23$ $4x = 32$ $x = 8$

2. 3x + 5 = x - 7 2x = -12 x = -6

3. 2x + 5 < 17 2x < 12 x < 6

Plotting Linear Equations and Calculating Slope

1. Slope = (6 - 2) ÷ (3 - (-1)) = 4 ÷ 4 = 1

2. Slope = (5 - (-3)) ÷ (4 - 0) = 8 ÷ 4 = 2

3. Slope = -2

Financial Math

1. Simple Interest: I = P x r x t I = $5,000 x 0.04 x 3 I = $600 You will earn $600 in interest.

2. Compound Interest (Annual): A = P(1 + r)^t A = $2,000(1 + 0.05)^4 A = $2,000(1.05)^4 A = $2,000 × 1.2155 A = $2,431.00 The final amount will be $2,431.00.

3. Compound Interest (Quarterly): A = P(1 + r÷n)^(nt) A = $3,000(1 + 0.06÷4)^(4 x 2) A = $3,000(1.015)^8 A = $3,000 x 1.1261 A = $3,378.30 The final amount will be $3,378.30.

Statistics and Probabilities

1. Dataset: 7, 9, 4, 6, 7, 10, 3, 7, 8, 5; Mean: (7+9+4+6+7+10+3+7+8+5)÷10 = 66/10 = 6.6 Median: Ordered set: 3, 4, 5, 6, 7, 7, 7, 8, 9, 10; Median = (6 + 7)÷2 = 6.5 Mode: 7 (occurs three times)

2. P(red or green) = P(red) + P(green) = 5/10 + 2/10 = 7/10 = 0.7

3. P(selecting 1, 2, and 3 in order) = 3/5 x 2/4 x 1/3 = 1/10 = 0.1

Reflection on Real-Life Math Applications

Financial Math

1. Simple Interest: $I = P \times r \times t$ $I = \$2,500 \times 0.045 \times 3 = \$337.50$

2. Compound Interest (Annual): $A = P(1 + r)^t$ $A = \$3,000(1 + 0.06)^4 = \$3,780.54$

3. Comparison: a) Simple Interest: $A = P(1 + rt) = \$5,000(1 + 0.05 \times 2) = \$5,500$ b) Compound Quarterly: $A = P(1 + r \div n)^{(nt)} = \$5,000(1 + 0.05 \div 4)^{(4 \times 2)} = \$5,519.64$ Compound interest yields $19.64 more.

Statistics and Probabilities

1. Dataset: 14, 18, 22, 14, 16, 18, 20, 22, 14, 18; Mean: $(14+18+22+14+16+18+20+22+14+18) \div 10 = 176/10 = 17.6$; Median: Ordered set: 14, 14, 14, 16, 18, 18, 18, 20, 22, 22; Median = $(18 + 18) \div 2 = 18$; Mode: 14 and 18 (both occur three times)

2. P(heart or king) = P(heart) + P(king) - P(heart and king) = $13/52 + 4/52 - 1/52 = 16/52 = 4/13$ b) P(red or ace) = P(red) + P(ace) - P(red and ace) = $26/52 + 4/52 - 2/52 = 28/52 = 7/13$

3. P(rain and wind) = P(rain) x P(wind) = $0.4 \times 0.3 = 0.12$

Practice Test

Numbers and Operations

1. $18 + 3 \times (7 - 2) \div 5 + 2^2$; Step 1: $(7 - 2) = 5$; Step 2: $2^2 = 4$, $18 + 3 \times 5/5 + 4$; Step 3: $3 \times 5 = 15$, $15/5 = 3$, $18 + 3 + 4$; Step 4: $18 + 3 = 21$, $21 + 4 = 25$; Final Answer: 25

2. 24/4 x 3 - 5 + 2^3; Step 1: 2^3 = 8, 24/4 x 3 - 5 + 8; Step 2: 24/4 = 6, 6 x 3 - 5 + 8; Step 3: 6 x 3 = 18, 18 - 5 + 8; Step 4: 18 - 5 = 13, 13 + 8 = 21; Final Answer: 21

3. 5 + 3 x (4 - 2)^2 - 6/2; Step 1: (4 - 2) = 2, 5 + 3 x 2^2 - 6/2; Step 2: 2^2 = 4, 5 + 3 x 4 - 6/2; Step 3: 3 x 4 = 12, 6/2 = 3, 5 + 12 - 3; Step 4: 5 + 12 = 17, 17 - 3 = 14; Final Answer: 14

Fractions and Decimals

1. 3/8 + 0.625; Step 1: Convert 3/8 to a decimal 3/8 = 0.375; Step 2: Add the decimals 0.375 + 0.625 = 1.000; Final Answer: 1.000 or simply 1

2. 2/3 x 0.75; Step 1: Convert 0.75 to a fraction 0.75 = 75/100 = 3/4; Step 2: Multiply the fractions 2/3 x 3/4 = 6/12 = 1/2; Final Answer: 1/2

3. Convert 5/6 to a decimal; Step 1: Divide 5 by 6 5/6 = 0.8333333...; Step 2: Round to three decimal places; Final Answer: 0.833

Percentages and Ratios

1. In a class of 200 students, 45% are boys. How many girls are in the class? Step 1: Calculate the number of boys 45% of 200 = 0.45 x 200 = 90 boys; Step 2: Subtract boys from total to get girls 200 - 90 = 110 girls; Final Answer: There are 110 girls in the class.

2. A shirt originally priced at $80 is on sale for 15% off. What is the sale price? Step 1: Calculate the discount amount 15% of $80 = 0.15 x $80 = $12; Step 2: Subtract the discount from the original price $80 - $12 = $68; Final Answer: The sale price is $68.

3. The ratio of cats to dogs in a pet store is 3:5. If there are 24 dogs, how many cats are there? Step 1: Set up the proportion 3/5 = x/24, where x is the number of cats; Step 2: Cross-multiply and solve for "x" 3 x 24 = 5x 72 = 5x

x = 72/5 = 14.4; Final Answer: There are 14 cats (rounded down as we can't have a fraction of a cat).

Factors and Multiples

1. Prime factorization of 84: 84/2 = 42, 42/2 = 21, 21/3 = 7, 7 is prime; therefore, 84 = 2 x 2 x 3 x 7 = 2^2 x 3 x 7

2. LCM of 18 and 24 using prime factorization: 18 = 2 x 3 x 3 = 2 x 3^2, 24 = 2 x 2 x 2 x 3 = 2^3 x 3; LCM = 2^3 x 3^2 = 8 x 9 = 72

3. Factors of 72 using prime factorization: 72 = 2^3 x 3^2; Factors: 1, 2, 3, 4, 6, 8, 9, 12, 18, 24, 36, 72 (All combinations of 2^0, 2^1, 2^2, 2^3 and 3^0, 3^1, 3^2)

Basic Algebraic Concepts

1. Simplify: 3x + 2y - 5x + 4y - 7; Step 1: Group like terms (3x - 5x) + (2y + 4y) - 7; Step 2: Combine like terms -2x + 6y - 7; Final Answer: -2x + 6y - 7

2. Combine like terms: 2a^2 - 3ab + 5a^2 + 2ab - 4a^2; Step 1: Group like terms (2a^2 + 5a^2 - 4a^2) + (-3ab + 2ab); Step 2: Combine like terms 3a^2 - ab; Final Answer: 3a^2 - ab

3. Simplify: 4(x + 2) - 2(x - 3) + 3x; Step 1: Distribute 4x + 8 - 2x + 6 + 3x; Step 2: Group like terms (4x - 2x + 3x) + (8 + 6); Step 3: Combine like terms 5x + 14; Final Answer: 5x + 14

Integers and Exponents

1. Evaluate: (-3) + 7 - (-5) + (-2); Step 1: Simplify -(-5) to +5 (-3) + 7 + 5 + (-2); Step 2: Add from left to right 4 + 5 + (-2) = 9 + (-2) = 7; Final Answer: 7

2. Multiply: $(-4) \times 3 \times (-2)$; Step 1: Multiply the first two numbers $(-4) \times 3 = -12$; Step 2: Multiply the result by the last number $-12 \times (-2) = 24$; Final Answer: 24

3. Simplify using the power of powers rule: $(x^3)^4$; Apply the rule: $(x^m)^n = x^{(m.n)}$ $(x^3)^4 = x^{3^4} = x^{12}$; Final Answer: x^{12}

Proportions and Rates

1. Recipe problem: Set up the proportion: $2/5 = x/8$, where x is cups of sugar Cross-multiply: $5x = 2 \times 8$ Solve for x: $5x = 16$ $x = 16 \div 5 = 3.2$ Final Answer: You need 3.2 cups of sugar.

2. Car travel problem: Set up the rate: 240 miles $\div$ 4 hours = 60 miles per hour; for 6 hours: 60 miles/hour $\times$ 6 hours = 360 miles; Final Answer: The car will travel 360 miles in 6 hours.

3. Teacher-student ratio problem: Set up the proportion: $1/25 = x/750$, where x is the number of teachers Cross-multiply: $25x = 1 \times 750$; Solve for x: $25x = 750$ $x = 750/25 = 30$ Final Answer: There are 30 teachers.

Geometry Fundamentals

1. Rectangular garden: Perimeter = 2(length + width) = 2(15 + 10) = 2(25) = 50 feet; Area = length x width = 15 x 10 = 150 square feet

2. Circular pond: Circumference = $2\pi r$ = 2 x 3.14 x 4 = 25.12 meters; Area = πr^2 = 3.14 x 4^2 = 3.14 x 16 = 50.24 square meters

3. Triangular flag: Perimeter = 8 + 5 + 5 = 18 inches; Area = (1/2) x base x height = (1/2) x 8 x 6 = 24 square inches

Shapes and Their Properties

1. Rectangular Prism: Volume = length x width x height V = 8 cm x 5 cm x 3 cm = 120 cubic centimeters (cm^3)

2. Cylindrical Water Tank: Volume = πr^2h; V = 3.14 x 2^2 x 5 V = 3.14 x 4 x 5 = 62.8 cubic meters (m^3)

3. Sphere: Volume = (4/3)πr^3 V = (4/3) x 3.14 x 6^3; V = (4/3) x 3.14 x 216 = 904.32 cubic inches (in^3)

Algebraic Expressions

1. Factor: 12x^2y - 18xy^2 + 24xy; Step 1: Find the greatest common factor, GCF = 6xy; Step 2: Factor out the GCF 6xy(2x - 3y + 4); Final answer: 6xy(2x - 3y + 4)

2. Factor: x^2 + 7x + 12; Step 1: Find two numbers that multiply to 12 and add to 7, 3 and 4 satisfy this condition; Step 2: Rewrite the middle term using these numbers x^2 + 3x + 4x + 12; Step 3: Factor by grouping x(x + 3) + 4(x + 3) (x + 3)(x + 4); Final answer: (x + 3)(x + 4)

3. Simplify: 3x^3 - 2x^2 + 5x - 4x^3 + 6x^2 - 2x + 8; Step 1: Group like terms (3x^3 - 4x^3) + (-2x^2 + 6x^2) + (5x - 2x) + 8; Step 2: Combine like terms -x^3 + 4x^2 + 3x + 8; Final answer: -x^3 + 4x^2 + 3x + 8

Linear Equations and Inequalities

1. Solve: 0.5x + 1.2 = 0.8x - 0.6; Step 1: Subtract 0.5x from both sides 1.2 = 0.3x - 0.6; Step 2: Add 0.6 to both sides 1.8 = 0.3x; Step 3: Divide both sides by 0.3 x = 1.8 ÷ 0.3 = 6 Final answer: x = 6

2. Solve: [2x - 5] = 11; This equation splits into two cases: Case 1: 2x - 5 = 11 2x = 16 x = 8; Case 2: 2x - 5 = -11 2x = -6 x = -3; Final answer: x = 8 or x = -3

3. Solve: (x + 2)÷(x - 3) ≤ 2; Step 1: Multiply both sides by (x - 3), considering two cases: Case 1 (x - 3 > 0): x + 2 ≤ 2(x - 3) x + 2 ≤ 2x - 6 8 ≤ x; Case 2 (x - 3 < 0): x + 2 ≥ 2(x - 3) x + 2 ≥ 2x - 6 8 ≥ x; Step 2: Combine the results, considering the domain x ≠ 3 Final answer: x ≥ 8 or x < 3

Graphing

1. Slope between points (-3, 4) and (2, -1): Use the slope formula: $m = (y_2 - y_1) \div (x_2 - x_1)$ $m = (-1 - 4) \div (2 - (-3))$ $m = -5 \div 5 = -1$; Answer: The slope is -1.

2. Slope of 3x - 2y = 12: Rearrange to slope-intercept form: y = mx + b 2y = 3x - 12 y = (3/2)x - 6 Answer: The slope is 3/2 or 1.5.

3. Roof slope: Slope = rise ÷ run = 5 feet ÷ 20 feet = 1/4 or 0.25 Answer: The slope is 1/4 or 0.25 (often expressed as "1 in 4" in construction).

Financial Math

1. I = P x r x t; I = $5,500 x 0.045 x 3, I = $742.50; The interest paid will be $742.50.

2. $A = P(1 + r \div n)^{(nt)}$; $A = 3000(1 + 0.06 \div 4)^{(4 \times 2)}$; $A = 3000(1.015)^8$; A = $3,385.80 The balance after 2 years will be $3,385.80.

3. Using the formula: t = 100 ÷ r (where t is time in years and r is the annual interest rate); t = 100 ÷ 8 = 12.5 years. It will take 12.5 years for the amount to double.

Statistics and Probabilities

1. Mean = (3+5+2+4+3+6+2+4+5)/9 = 34/9 = 3.78 Median (ordered data): 2, 2, 3, 3, 4, 4, 5, 5, 6. Median = 4; Mode: 3, 4, and 5 all appear twice. This dataset is trimodal.

2. P(1st green) = 5/12 P(2nd green, given 1st was green) = 4/11 P(both green) = 5/12 x 4/11 = 20/132 = 5/33 ≈ 0.1515 or about 15.15%

3. Mean = (12+15+11+13+15+14+12)/7 = 92/7 = 13.14 kg. For standard deviation: $\Sigma(x - mean)^2 = (12-13.14)^2 + (15-13.14)^2 + ... + (12-13.14)^2 = 20.86$. Standard Deviation = $\sqrt{(20.86/7)} = \sqrt{2.98} = 1.73$ kg

---

References

BBC. (n.d.). *Perimeter - 2-dimensional shapes - edexcel - GCSE maths revision - edexcel*. BBC Bitesize. https://www.bbc.co.uk/bitesize/guides/zgbd2nb/revision/2

Bhandari, P. (2020, July 30). *Central tendency | understanding the mean, median and mode*. Scribbr. https://www.scribbr.com/statistics/central-tendency/

Blankman, R. (2021, February 8). *Teaching ratios and unit rates in math*. Hmhco. https://www.hmhco.com/blog/teaching-ratios-and-unit-rates-in-math

Brack, T. (2022, October 4). *Simplifying expressions by combining like terms*. Maneuvering the Middle. https://www.maneuveringthemiddle.com/tips-for-teaching-simplifying-expressions/

de la Fuente, P. (2022, January 28). *50/30/20 budget calculator*. NerdWallet. https://www.nerdwallet.com/article/finance/nerdwallet-budget-calculator

Financial ratios: 4 ways to assess your business. (2020, September 12). BDC. https://www.bdc.ca/en/articles-tools/money-finance/manage-finances/financial-ratios-4-ways-assess-business

Fractions & decimals: Real world applications - lesson. (2024). SStudy. https://study.com/academy/lesson/fractions-decimals-real-world-applications.html

General business management applications. (2014). NSCC. https://pressbooks.nscc.ca/bookkeepingmath/chapter/chapter-2-general-business-management-applications/

Hayes, A. (2019). *What modes mean in statistics*. Investopedia. https://www.investopedia.com/terms/m/mode.asp

Kenton, W. (2019). *Percentage change*. Investopedia. https://www.investopedia.com/terms/p/percentage-change.asp

Kit, M. (2018, October 9). *Thinking through LCM & GCF*. Teachdomore. https://teachdomore.wordpress.com/2018/10/09/thinking-through-lcm-gcf-inspired-by-pwharris-visit/

Lake, R. (2023, October 26). *50/30/20 budget calculator*. Forbes. https://www.forbes.com/advisor/banking/budget-calculator/

Library and Learning Centre. (2024). *Library guides: Math help from the learning centre: Basics of graphing*. Centennialcollege.ca. https://libraryguides.centennialcollege.ca/c.php?g=645085&p=5132710

Like terms made simple: Simplification techniques. (2021, March 3). Litutorl. https://iitutor.com/simplifying-multiple-like-terms/

Linear inequalities questions. (2024, February 19). Third Space Learning. https://thirdspacelearning.com/gcse-maths/algebra/linear-inequalities/

Lynch, C. (2022, March 2). *Area and perimeter of 2D shapes*. PlanBee. https://planbee.com/blogs/news/area-and-perimeter-of-2d-shapes

Mathematical numbers: Natural, whole, rational, irrational, real, complex, integers. (2020). UniversalClass. https://www.universalclass.com/articles/math/pre-calculus/natural-whole-rational-irrational-real-complex-integers.htm

Milne, C. (2020). *Angles unveiled: Exploring their practical applications in real life*. Abakus Europe. https://abakus-center.com/blog/angles-unveiled

Number systems: Naturals, integers, rationals, irrationals, reals, and beyond. (n.d.). Varsitytutors. https://www.varsitytutors.com/hotmath/hotmath_help/topics/number-systems

Polynomials: Mastering polynomials with the algebraic method. (2024, June 18). FasterCapital. https://fastercapital.com/content/Polynomials--Mastering-Polynomials-with-the-Algebraic-Method.html

Resource Center. (n.d.). *Order of operations*. Content.byui.edu. https://content.byui.edu/file/b8b83119-9acc-4a7b-bc84-efacf9043998/1/Math-1-6-1.html

Types of angles | learn with real-life examples. (n.d.). Tutoringhour. https://www.tutoringhour.com/lessons/geometry/types-of-angles/

What is order of operations? - definition, facts & example. (n.d.). Splashlearn. https://www.splashlearn.com/math-vocabulary/algebra/order-of-operations

Whiteside, E. (2022, September 17). *What is the 50/20/30 budget rule?* Investopedia. https://www.investopedia.com/ask/answers/022916/what-502030-budget-rule.asp

Wright, J. (2024, June 10). *Library guides: Math skills overview guide: Number sets.* Davenport.libguides.com. https://davenport.libguides.com/math-skills-overview/basic-operations/sets

Made in the USA
Las Vegas, NV
30 November 2024

12947279R00079